ANTIEIGENVALUE ANALYSIS

With Applications to Numerical Analysis, Wavelets, Statistics, Quantum Mechanics, Finance and Optimization

ANTIEIGENVALUE ANALYSIS

With Applications to Numerical Analysis, Wavelets,
Statistics, Quantum Mechanics, Finance and Optimization

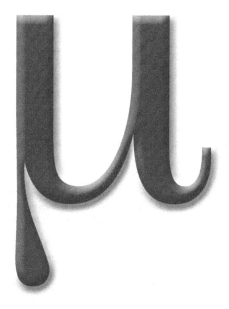

KARL GUSTAFSON

University of Colorado at Boulder, USA

World Scientific

NEW JERSEY · LONDON · SINGAPORE · BEIJING · SHANGHAI · HONG KONG · TAIPEI · CHENNAI

Published by

World Scientific Publishing Co. Pte. Ltd.

5 Toh Tuck Link, Singapore 596224

USA office: 27 Warren Street, Suite 401-402, Hackensack, NJ 07601

UK office: 57 Shelton Street, Covent Garden, London WC2H 9HE

British Library Cataloguing-in-Publication Data
A catalogue record for this book is available from the British Library.

Jacket photograph of the author by Jillian Lloyd

For photocopying of material in this volume, please pay a copying fee through the Copyright Clearance Center, Inc., 222 Rosewood Drive, Danvers, MA 01923, USA. In this case permission to photocopy is not required from the publisher.

ISBN-13 978-981-4366-28-1
ISBN-10 981-4366-28-5

Typeset by Stallion Press
Email: enquiries@stallionpress.com

Printed in Singapore.

The senses are said to be higher than the body,
the mind higher than the senses;
above the mind is the intellect,
and above the intellect is the Ātman.
The Bhagavad Gita

To my grandchildren
Ashley, Elizabeth, Francesca, Clarissa, and Julian

Preface

In 1966 I published my first mathematical journal paper [Gustafson (1966)] and shortly after that I published the related papers [Gustafson (1967; 1968a, 1968b, 1968c, 1968d)]. These papers contain, although by other names, the essential ingredients of what became my theory of antieigenvalues. In my first plenary lecture, at the *Third Symposium on Inequalities* held at the University of California in Los Angeles, September 1–9, 1969, I specifically introduced the term *antieigenvalue*, mentioned my Euler equation for them, and proved my Min-Max Theorem, a result fundamental to my establishing the antieigenvalue full trigonometric theory. This lecture appeared in the rather short paper [Gustafson (1972a)].

But after 1970 I was off to other mathematical pursuits, notably, quantum scattering theory, then statistical mechanics, and then computational fluid dynamics. Except for a couple of papers on the antieigenvalue theory during the next two decades with my Ph.D. students Duggirala Rao and Morteza Seddighin, I had left behind my theory of antieigenvalues. In retrospect, I had not yet pushed it into the light of day.

I was brought back to further develop my antieigenvalue theory after 1990 by an invitation to give a plenary lecture at the *Fourth International Workshop in Analysis and its Applications* held in Dubrovnik, Yugoslavia, June 1–10, 1990. I was surprised that the organizers wanted me to speak about my results of more than twenty years earlier. But then I saw that the principal theme of that conference was: Inner Product and Convexity Structures in Analysis, Mathematical Physics and Economics. Indeed, my antieigenvalue theory naturally falls within those structures and application areas. My 1990 lecture in Yugoslavia appeared a year later in [Gustafson (1991a)]. In that paper, after bringing my theory up-to-date, I established the new result that the important Kantorovich bound for quadratic optimizaton in economics is fundamentally trigonometric, and moreover in a way seen quite quickly and explicitly within my antieigenvalue theory. Later I would apply my antieigenvalue theory to clarify the Bell inequalities in mathematical physics. So the Dubrovnik invitation was useful: indeed prescient.

In 1995 I gave a series of invited lectures in Japan and those resulted in the book *Lectures in Computational Fluid Dynamics, Mathematical Physics, and Linear Algebra* [see Gustafson (1996a; 1997d)]. Part III of that book presents my antieigenvalue theory as known up to 1995. In those books and in my papers after 1990, I often present my new results for the antieigenvalue theory under the broader term *operator trigonometry*. Then in 1997 with Duggirala Rao I published another book, *Numerical Range: The Field of Values of Linear Operators and Matrices* [Gustafson (1997a)]. Chapter 3 there also presents the operator trigonometry as understood up to 1995.

The fifteen years since 1995 has seen a much wider range of applications of the antieigenvalue theory, accompanied by some generalizations and in particular some deeper understandings of it. The purpose of this book is to provide the first comprehensive treatment of the antieigenvalue theory and its applications.

Karl Gustafson
Boulder, Colorado

Contents

Preface   ix

1. Introduction   1

Perspective   1

1.1   A Recent Referee Speaks   2
1.2   The Original Motivation   2
1.3   The Essential Entities   3
1.4   Simple Examples and a Picture   6
1.5   Applications to-Date   9
1.6   Organization of this Book   10

Commentary   12

1.7   Exercises   14

2. The Original Motivation: Operator Semigroups   15

Perspective   15

2.1   Abstract Initial Value Problems   16
2.2   The Hille–Yosida–Phillips–Lumer Theorem   17
2.3   The Rellich–Kato–Nelson–Gustafson Theorem   17
2.4   The Multiplicative Perturbation Theorem   18
2.5   When are Positive Operator Products Positive?   20
2.6   Nonnegative Contraction Semigroups   21

Commentary   **22**

2.7   Exercises   23

3.   The Essentials of Antieigenvalue Theory   **25**

Perspective   **25**

3.1   Convexity Properties of Norm Geometry   26
3.2   The Min-Max Theorem   27
3.3   The Euler Equation   33
3.4   Higher Antieigenvalues and Antieigenvectors   39
3.5   The Triangle Inequality   46
3.6   Extended Operator Trigonometry   47

Commentary   **49**

3.7   Exercises   50

4.   Applications in Numerical Analysis   **53**

Perspective   **53**

4.1   Gradient Descent: Kantorovich Bound is Trigonometric . . .   54
4.2   Minimum Residual $Ax = b$ Solvers   56
4.3   Richardson Relaxation Schemes (e.g. SOR)   57
4.4   Very Rich Trigonometry Underlies ADI   60
4.5   Domain Decomposition Multilevel Schemes   61
4.6   Preconditioning and Condition Numbers   63

Commentary   **65**

4.7   Exercises   67

5.   Applications in Wavelets, Control, Scattering   **69**

Perspective   **69**

5.1   The Time Operator of Wavelets   70
5.2   Frame Operator Trigonometry   74
5.3   Wavelet Reconstruction is Trigonometric   76
5.4   New Basis Trigonometry   78

5.5 Trigonometry of Lyapunov Stability 85

5.6 Multiplicative Perturbation and Irreversibility 86

Commentary 88

5.7 Exercises 89

6. **The Trigonometry of Matrix Statistics** 91

Perspective 91

6.1 Statistical Efficiency 91

6.2 The Euler Equation versus the Inefficiency Equation 101

6.3 Canonical Correlations and Rayleigh Quotients 105

6.4 Other Statistics Inequalities 107

6.5 Prediction Theory: Association Measures 112

6.6 Antieigenmatrices 115

Commentary 116

6.7 Exercises 118

7. **Quantum Trigonometry** 123

Perspective 123

7.1 Bell–Wigner–CHSH Inequalities 125

7.2 Trigonometric Quantum Spin Identities 129

7.3 Quantum Computing: Phase Issues 132

7.4 Penrose Twistors 135

7.5 Elementary Particles 142

7.6 Trigonometry of Quantum States 144

Commentary 152

7.7 Exercises 153

8. **Financial Instruments** 155

Perspective 155

8.1 Some Remarks on Mathematical Finance 161

8.2 Quantos: Currency Options 167
8.3 Multi-Asset Pricing: Spread Options 172
8.4 Portfolio Rebalancing 175
8.5 American Options with Random Volatility 177
8.6 Risk Measures for Incomplete Markets 179

Commentary **181**

8.7 Exercises 182

9. Other Directions **183**

Perspective **183**

9.1 Operators 183
9.2 Angles 186
9.3 Optimization 189
9.4 Equalities 191
9.5 Geometry 194
9.6 Applications 199

Commentary **200**

9.7 Exercises 200

Appendix A   Linear Algebra **203**

A.1 Matrix Analysis 203
A.2 Operator Theory 204

Appendix B   Hints and Answers to Exercises **205**

Bibliography **229**

Index **241**

Chapter 1
Introduction

Perspective

In mathematics, engineering, and science, *eigenvalues* λ and their corresponding *eigenvectors* x occur in all linear systems. For a matrix A, they are the particular nonzero vectors x and scalars λ such that $Ax = \lambda x$. In other words, the vector direction of x is preserved under the action of the matrix A, subject to a scale change λ. The same situation occurs for A being a linear differential operator. There the eigenvalues and eigenvectors may be the characteristic vibration frequencies and the fundamental modes of the whole system. Much of science can be accomplished by writing an arbitrary vector as a linear combination of the eigenvectors of the relevant system, from which the behavior of the whole system can be seen in terms of that of its fundamental parts. Fourier analysis is exactly this. The more general theory is called spectral theory.

About forty years ago I was led from a question about partial differential equations to formulate the notions of *antieigenvalues* μ and their corresponding *antieigenvectors* x. Antieigenvectors x are those vectors most turned by a matrix or operator A. The corresponding antieigenvalue μ is the cosine of that maximal turning angle. The maximal turning angle I called $\phi(A)$, the angle of the operator. Just as you can order the eigenvalues λ from smallest to largest, my theory of antieigenvalues also creates an increasing sequence of antieigenvalues and their corresponding decreasing critical *operator turning angles* accompanied by their attendant (higher) antieigenvectors.

This first chapter will give you an overview of the whole subject, and of the whole book.

1.1 A Recent Referee Speaks

It is amusing yet extremely useful to begin with a direct quote taken from a
very conscientious referee of my recent paper [Gustafson (2009b)]. This was
one of the best and detailed referee reports that I have ever been privileged
to receive. I do not know who the referee was, but it was an expert in
operator theory who apparently had not previously been familiar with my
antieigenvalue theory. "In the late sixties Karl Gustafson, motivated by a
problem in perturbation theory of semigroups, noted the following very
interesting fact: Let A be a real positive definite $n \times n$ matrix on $\mathbb{R}^n$. Then
the two quantities

$$\inf_{x \neq 0} \frac{\langle Ax, x \rangle}{\|x\| \|Ax\|}, \quad \text{and} \quad \inf_{\epsilon > 0} \sup_{\|x\| \leq 1} \|(\epsilon A - I)x\|$$

satisfy the same relation as the cosine and sine (respectively) of an angle:
the sum of their squares yields 1."

The referee had honed in on the fact that for such matrices A with
antieigenvalues $0 < \lambda_1 \overset{\leq}{=} \lambda_2 \overset{\leq}{=} \cdots \overset{\leq}{=} \lambda_n$, I had shown in 1968 that the left
and right quantities above have values

$$\mu_1 = \frac{2\sqrt{\lambda_1 \lambda_n}}{\lambda_1 + \lambda_n} \quad \text{and} \quad \nu_1 = \frac{\lambda_n - \lambda_1}{\lambda_1 + \lambda_n}.$$

Clearly $\mu_1^2 + \nu_1^2 = 1$. I called μ_1 the first antieigenvalue of A. Then setting
$\cos \phi(A) = \mu_1$ determines the largest angle $\phi(A)$ through which A may turn
any vector. The angle $\phi(A)$ I called the angle of A.

I liked the referee's perception of my operator trigonometry because
it nicely brings out the two key elements of my theory for the special case
of $n \times n$ symmetric matrices A. However, that is not how I stumbled upon
my antieigenvalue theory. So let us get to the source.

1.2 The Original Motivation

Antieigenvalue theory and its main features were created by this author
in the period 1966–1969. The inducing application was the contraction
semigroup theory of abstract initial value problems

$$\begin{cases} \dfrac{du}{dt} = Au, & t > 0, \\[2mm] u(0) = u_0 & \text{given.} \end{cases}$$

For example, you may think of A as a symmetric matrix with all of its eigenvalues λ_i negative. Then the solution is

$$u(t) = e^{tA}u_0$$

and this solution has all modes decaying. The same initial value problem with A being a partial differential operator may be treated in the same way, using methods of functional analysis to exponentiate. There, and in the more general abstract semigroup theory, one just writes $u(t) = Z_t u_0$ and says that Z_t is a contraction semigroup with the nongrowing norm property $\|Z_t\| \overset{\leq}{=} 1$ in the appropriate Banach or Hilbert space.

I wanted to know when I could multiplicatively perturb the infinitesimal generator A by left multiplication by another operator B such that the initial value problem

$$\begin{cases} \dfrac{du}{dt} = BAu, & t > 0, \\ u(0) = u_0 & \text{given} \end{cases}$$

would still exponentiate to a contraction semigroup. In terms of the simple example where A and B both are $n \times n$ symmetric matrices, we may intuitively say: well, A is negative, we want BA to be negative, so let us take B to be positive. The fact that life is not that simple led me to formulate my antieigenvalue theory.

I will show in Chapter 2 more particulars of this originating question and exactly how it, in a very fortuitous fashion, brought forth two of the key elements of what became the antieigenvalue theory. But in this introductory chapter I want to be sure that the reader may understand in simplified terms the basic concepts of the antieigenvalue theory. Let us get to those now.

1.3   The Essential Entities

For clarity of presentation we now take both A and B to be "positive": so in the semigroup perturbation above, it will be $-A$ and $-BA$ which will exponentiate to a contraction semigroup. We may also stay mentally in the special case of A and B both symmetric positive definite matrices which thus have all of their eigenvalues strictly positive. If B commutes with A, then it is easy to show that BA is still symmetric positive definite and $-BA$ will generate a contraction semigroup. One may similarly treat more general commuting operators A and B. For that reason, in 1967 I took the position that commuting operators are the trivial case, and focused on the general noncommuting case.

In that general situation, what is needed for $-BA$ to generate a contraction semigroup is that BA be maximally accretive: all of its spectrum $\sigma(BA)$ must be in the closed right half-plane and so must the numerical range $W(BA)$ of its quadratic form values $\langle BAx, x \rangle$ be. See the books [Hille and Phillips (1957); Kato (1976)] for more particulars.

Going back to our noncommuting matrices A and B, what is needed is that the real part of BA be nonnegative. Recall that BA will not be symmetric unless B and A commute and we have discarded that situation as a trivial case. For such matrix products, their (symmetric) real part is

$$\operatorname{Re} BA = \frac{BA + (BA)^*}{2} = \frac{BA + AB}{2},$$

with the latter equality valid in our special case of $A = A^*$ and $B = B^*$.

For the general case in Banach space I established the sufficient condition

$$\inf_{\epsilon > 0} \sup_{\|x\| \leq 1} \|(\epsilon B - I)x\| \leq \inf_{x \neq 0} \frac{\operatorname{Re} \langle Ax, x \rangle}{\|Ax\| \|x\|} \qquad (*)$$

for $-BA$ to remain a contraction semigroup generator. I did not like the implicit restriction for a bounded B and later somewhat extended the criteria to unbounded multiplicative perturbations. But for now and in most of this book, let us take B on the left-hand side of the inequality (*) to be everywhere defined and bounded on a Hilbert space X. Also the right-hand side of the inequality (*) needs to be stipulated as pertaining only for x in the domain $D(A)$ of the generally unbounded operator A. By the way, in a Banach space the entity $\langle y, x \rangle$ denotes a semi-inner product, of which each Banach space possesses at least one. But most of this book will be in Hilbert space.

Let us leave those technicalities aside and bounce back to our symmetric positive definite matrices A and B. Then the (*) inequality above becomes

$$\min_{\epsilon > 0} \|\epsilon B - I\| \leq \min_{x \neq 0} \frac{\langle Ax, x \rangle}{\|Ax\| \|x\|}.$$

I thought it natural to call the right-hand side of this general inequality $\cos \phi(A)$ and to name it the first (real) antieigenvalue μ_1 of A. But how about the left-hand side?

For B a bounded strongly accretive operator on a Banach space, i.e., Re $\langle Bx, x \rangle \overset{\geq}{=} m_B > 0$ for all $\|x\| = 1$, it is a fact that the convex norm curve $\|\epsilon B - I\|$ dips below 1 for some positive interval $(0, \epsilon_1)$. Moreover for B on a Hilbert space, there will be a unique ϵ_m for which the norm curve $\|\epsilon B - I\|$ achieves its minimum. So the above key $(*)$ inequality becomes

$$\nu_1 \equiv \|\epsilon_m B - I\| \overset{\leq}{=} \mu_1 \equiv \cos \phi(A).$$

I wanted the left-hand side also to be trigonometric and managed to prove the following fundamental result:

Theorem 1.1 (Min-Max Theorem). *Let A be a strongly accretive bounded operator on a Hilbert space. Then*

$$\sup_{\|x\| \overset{\leq}{=} 1} \inf_{-\infty < \epsilon < \infty} \|(\epsilon A - I)x\|^2 = \inf_{\epsilon > 0} \sup_{\|x\| \overset{\leq}{=} 1} \|(\epsilon A - I)x\|^2.$$

The right-hand side is $\|\epsilon_m A - I\|^2$ and one can show that the left-hand side is $1 - \cos^2 \phi(A)$. With this fundamental theorem I had an operator trigonometry. That is, I was now entitled to name

$$\nu_1 \equiv \|\epsilon_m B - I\| \equiv \sin \phi(B).$$

You do not have a trigonometry unless you have both a cosine and a sine.

Thus I arrived at the operator-trigonometric sufficient condition for $-BA$ to generate a contraction semigroup:

$$\sin \phi(B) \overset{\leq}{=} \cos \phi(A).$$

For the special case of our symmetric positive definite (SPD) matrices A and B, it came out of my proof of the Min-Max Theorem that

$$\cos \phi(A) = \frac{2\sqrt{\lambda_1 \lambda_n}}{\lambda_1 + \lambda_n}, \quad \sin \phi(A) = \frac{\lambda_n - \lambda_1}{\lambda_n + \lambda_1}.$$

Here I have ordered the eigenvalues

$$0 < \lambda_1 \overset{\leq}{=} \lambda_2 \overset{\leq}{=} \cdots \overset{\leq}{=} \lambda_n$$

as is my, and the operator theory community's, custom. Moreover I was able to extract from my proof of the Min-Max Theorem that the most-turned vectors for an SPD matrix A come as the pair

$$x_{\pm} = \pm \left(\frac{\lambda_n}{\lambda_1 + \lambda_n}\right)^{1/2} x_1 + \left(\frac{\lambda_1}{\lambda_1 + \lambda_n}\right)^{1/2} x_n.$$

Here x_1 and x_n are any norm-one eigenvectors from the λ_1 and λ_n eigenspaces, respectively.

In a similar way, one can define higher antieigenvalues μ_i and corresponding (smaller) turning angles $\phi_i(A)$ in terms of the intermediate eigenvectors. See Example 1.3.

1.4 Simple Examples and a Picture

What are those operator-turning angles ϕ_i and the antieigenvalues μ_i? They are best first understood by simple examples. Here are three.

Example 1.1.

$$A = \begin{bmatrix} 1 & 0 \\ 0 & 2 \end{bmatrix},$$

$$\cos\phi_1(A) = \frac{2\sqrt{\lambda_1 \lambda_2}}{\lambda_1 + \lambda_2} = \frac{2\sqrt{2}}{3} \cong 0.94281,$$

$$\phi_1(A) = 19.417 \text{ degrees}$$

Example 1.2.

$$A = \begin{bmatrix} 9 & 0 \\ 0 & 16 \end{bmatrix},$$

$$\cos\phi_1(A) = \frac{2\sqrt{\lambda_1 \lambda_2}}{\lambda_1 + \lambda_2} = \frac{2\sqrt{144}}{25} = 0.96,$$

$$\phi_1(A) = 16.269 \text{ degrees}.$$

Example 1.3.

$$A = \begin{bmatrix} 1 & 0 & 0 & 0 \\ 0 & 2 & 0 & 0 \\ 0 & 0 & 10 & 0 \\ 0 & 0 & 0 & 20 \end{bmatrix},$$

$$\cos\phi_1(A) = \frac{2\sqrt{\lambda_1\lambda_4}}{\lambda_1 + \lambda_4} = \frac{2\sqrt{20}}{21} = 0.42591771,$$

$$\phi_1(A) = 64.7912347 \text{ degrees,}$$

$$\cos\phi_2(A) = \frac{2\sqrt{\lambda_2\lambda_3}}{\lambda_2 + \lambda_3} = \frac{2\sqrt{20}}{12} = 0.74535593,$$

$$\phi_2(A) = 41.8103149 \text{ degrees.}$$

The antieigenvectors are easily calculated from the above formula. For Example 1.1 they are

$$x_\pm = \frac{1}{\sqrt{3}}\begin{bmatrix} \pm\sqrt{2} \\ 1 \end{bmatrix};$$

for Example 1.2 they are

$$x_\pm = \frac{1}{5}\begin{bmatrix} \pm 4 \\ 3 \end{bmatrix};$$

for Example 1.3 the first antieigenvector pair is

$$x_\pm^1 = \frac{1}{\sqrt{21}}\begin{bmatrix} \pm\sqrt{20} \\ 0 \\ 0 \\ 1 \end{bmatrix}.$$

Then for this third example one may strip off the lowest and highest eigenspaces which determine the most turned first antieigenvector pair and speak of the next (or interior) critical turning angle ϕ_2 determined by the submatrix

$$A_{23} = \begin{bmatrix} 2 & 0 \\ 0 & 10 \end{bmatrix},$$

from which the corresponding interior, or second antieigenvector pair for the original matrix A is computed as above to be

$$x_\pm^2 = \frac{1}{\sqrt{6}}\begin{bmatrix} 0 \\ \pm\sqrt{5} \\ 1 \\ 0 \end{bmatrix}.$$

Generally and as in the above, we like to normalize all the eigenvectors to length = norm = 1, and then the antieigenvectors also will have norm 1.

It has been said that "a picture is worth a thousand words." Let us look more closely at Example 1.1 above. Figure 1.1 conveniently summarizes all the key information. Notice that for a 2×2 SPD matrix the norm curve $\|\epsilon A - I\|$ for $\epsilon \geq 0$ is very simple. It is first the line

$$\ell_1 : \|(\epsilon A - I)x_n\| = 1 - \epsilon \lambda_1, \quad \epsilon \leq \frac{2}{\lambda_1 + \lambda_n},$$

and then the line

$$\ell_2 : \|(\epsilon A - I)x_1\| = -1 + \epsilon \lambda_n, \quad \epsilon \geq \frac{2}{\lambda_1 + \lambda_n}.$$

For our Example 1.1, these are just the lines

$$\ell_1 : 1 - \epsilon \quad \text{for } \epsilon \overset{\leq}{=} \frac{2}{3}$$

and the line

$$\ell_2 : -1 + 2\epsilon \quad \text{for } \epsilon \overset{\geq}{=} \frac{2}{3}.$$

The intersection of those two lines comes at $\epsilon_m = 2/3$ and the height of that intersection consists of $1/3 = \sin \phi(A)$. In other words, the norm curve $\|\epsilon A - I\|$ consists of a "one-component" line to the left, a "one-component" line to the right, and only where the lines intersect does it become a "two-component." In connection with Fig. 1.1 we note that

$$x\pm = \begin{bmatrix} \pm(2/3)^{1/2} \\ (1/3)^{1/2} \end{bmatrix} = \begin{bmatrix} 0.816496581 \\ 0.577350269 \end{bmatrix},$$

$$\frac{\langle Ax_+, x_+ \rangle}{\|Ax_+\| \|x_+\|} = \frac{2/3 + 2(1/3)}{(2)^{1/2}1} = \frac{2\sqrt{2}}{3}.$$

For general matrices and operators we do not know the exact antieigenvectors. Even for small normal matrices an antieigenvector may be one-component. In the most important applications, we are dealing with selfadjoint operators $A = A^*$. Then the picture of antieigenvectors as composed of exactly two components (or approximately two components in those infinite-dimensional cases in which spectral endpoints are not achieved by eigenvectors) is correct.

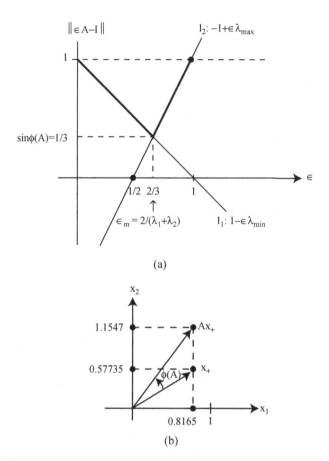

Fig. 1.1. Operator trigonometry of Example 1.1.
(a) The geometry of the curve $\|\epsilon A - I\|$ and $\sin\phi(A)$.
(b) The action of the antieigenvector x_+ under A.

1.5 Applications to-Date

Here, in summary form, are the main applications of the antieigenvalue theory and operator trigonometry to-date. These subjects have been arranged in roughly the chronological order in which I have treated them.

(1) Operator semigroups, differential equations
(2) Positivity of matrix products
(3) Markov processes
(4) Rayleigh–Ritz theory

(5) Numerical range

(6) Normal operators

(7) Convexity and inequality theory

(8) Minimum residual $Ax = b$ gradient solvers

(9) Richardson relaxation schemes (e.g., SOR)

(10) Domain decomposition multilevel schemes

(11) Preconditioning and condition numbers

(12) Wavelets

(13) Control theory

(14) Scattering theory

(15) Statistical estimation, efficiency, correlations

(16) Quantum Bell's inequalities

(17) Quantum computing

(18) Elementary particles and Penrose twistors

(19) Quantum entanglement and geometry of states

(20) Financial instruments

All of these applications will be discussed within this book, some more than others. It is my hope that this book will open up the subject and methods of antieigenvalue analysis and operator trigonometry to the wider world and thereby engender an enlarged range of application areas.

1.6   Organization of this Book

The Preface presented you the history of how I created, and then developed in intervals during my scientific career, antieigenvalue analysis and operator trigonometry. In this introductory chapter I provided you a simplified overview and the flavor of the whole subject. In Chapter 2 I will return to the originating application, the perturbation theory of operator semigroups. For this a background of partial differential equations and functional analysis is assumed. However, you can easily follow through by thinking in terms of your first courses in matrix analysis (alternatively, linear algebra or linear systems theory) and ordinary differential equations. Chapter 2 will contain some discussion of application areas (1)–(3) above.

Chapter 3 presents the essential elements of the antieigenvalue theory. Although I originally developed that theory thinking within the generality of Banach spaces, the theory is much nicer in Hilbert space and that is where most of this book will take place. The contexts of Chapter 3 will be those of application areas (4)–(7) above. In addition to deriving the Min-Max Theorem stated above and treating all of its ramifications, the variational theory of antieigenvalues is developed, and in particular their Euler equation is derived. Thus Chapters 2 and 3 reflect, roughly, my research results of the period 1966–1969.

Chapter 4 turns to my application of the antieigenvalue theory to numerical analysis. This came about from my experience acquired in research in computational fluid dynamics and other engineering projects in the 1980s. Throughout that decade I worked with several engineering groups and indeed produced Ph.D.'s in aeronautical, civil, chemical, and electrical engineering. These were great students from whom I learned a lot. Also I gained a steadily increasing appreciation for computational linear algebra, the guts of 70% of all scientific computing. However, I was so busy with the specific research tasking through large NSF grants with Boeing and our NSF Research Center for Optoelectronic Computing Systems that I had no time to set aside for my antieigenvalue theory. Thus the investigations and new results reported in Chapter 4 commenced shortly afterward, in the early 1990s. My interests in the operator trigonometry of those important Linear Solvers of application areas (8)–(11) continue to this day.

Chapter 5 could be thought of as miscellaneous applications. Although of course each of the subjects treated there carry their own importance, my applications of the antieigenvalue theory to wavelets, control, and scattering came about only intermittently, even, one could say, by chance. Because I carry my antieigenvalues and antieigenvectors within my subconscious, whenever I actively enter a new mathematical research arena, it is quite possible that eventually an opportunity to give to that area its new operator trigonometry will be too hard to resist. So Chapter 5 reflects *ad hoc* research results, in areas (12)–(14), sporadically obtained and presented in a few papers and at a few conferences during the last fifteen years, say, in the period since 1995.

I had known since the early days of my antieigenvalue theory that there would be interesting applications to matrix statistics. I even remarked upon this in [Gustafson (1972a)]. However, I did not get around to that until 1999 when I wrote the ground-breaking paper [Gustafson (1999e)]. That preprint was rejected by a journal but was consequently and subsequently circulated and it engendered considerable collateral interest within the matrix statistics community. In Chapter 6 I give the trigonometry of matrix statistics as I have developed it since 1999. Those results are all lumped into the applications area (15) above.

Because of my life-long interests in quantum mechanics it was inevitable to be drawn into giving new trigonometric interpretations to that large scientific enterprise, using the viewpoint of my antieigenvalue theory. Somewhat surprising in retrospect, they were issues in hidden variables and notably Bell's inequalities that created a natural entry point for me to put my operator trigonometry into quantum mechanics. Chapter 7 covers quantum mechanics application areas (16)–(19) and my work therein after my first paper [Gustafson (1999a)] up to the present.

Chapter 8 presents my very recent introduction of my operator trigonometry into the mathematics of high finance: applications area (20). I have been interested, even involved through teaching and research, in mathematical finance the last twenty years, but more along the lines of financial derivatives, notably the Black–Scholes partial differential equations and the variations and modernizations thereof. The other principal mathematical entity entering financial engineering is matrices. The topics treated in Chapter 8 use both partial differential equations and matrices. However, the emphasis there will be on new matrix trigonometry for finance.

Chapter 9 looks at other directions in which I or others have developed the operator trigonometry. Among those are antieigenvectors for normal operators, slant antieigenvalues, optimization theory, a proposed geometrical invariants theory, and motion analysis. One might call this chapter one of generalizations and incomplete future directions.

Commentary

This book is a research monograph. However, because my goal is to open up the methods of antieigenvalue analysis to a wider audience, I will often defer to the bibliographical literature some of the more involved technicalities, especially if they seem to be of narrow interest. So I found myself taking a guidebook kind of approach to this writing task, to which I could not allot more than one year's time. My belief that this was the best way to go was reinforced when I started writing the applications chapters. Substantial background is needed for a true understanding of each of those subjects.

In fact there were moments when I thought, why not just publish a compendium of all my papers that are on, or related to, my development of the antieigenvalue theory? As the Bibliography reveals, there are 75 of those. But that would not open the subject, although it might be useful to do so at a later date.

Although the antieigenvalue theory and evolving operator trigonometry are largely my creation, others have contributed. These include two of my Ph.D. students, Duggirala Rao and Morteza Seddighin. From the outside there were M. G. Krein and H. Wielandt who had early independent ideas similar to my operator cosine. But they did not see my operator sine and thus could not continue into a whole theory. G. Strang and J. Stampfli had early results that tangentially touch some of the inequalities of my antieigenvalue theory. C. Davis and B. Mirman obtained related results in the 1980s. Following my 1999 preprint on the operator trigonometry of matrix statistics, C. R. Rao and R. Khatree have given their versions, Rao in particular taking a useful wider view. All of these outside contributions I will detail later in the book.

Although this is a monograph, I decided it would be useful to append a few exercises at the end of each chapter. These are either easy or hard. The easy ones are meant to be reminders of elementary prerequisite facts. The harder ones might refer to the published literature. Some answers and hints are given at the back of the book in Appendix B. Appendices A.1 and A.2 contain extremely brief sketches of matrix analysis and operator theory, respectively. A few good books this author likes are cited there.

The Bibliography broke naturally into three separate parts. The first part Bib.1 gives my contributions (sometimes with coauthors). The second part Bib.2 lists important contributions to the theory by others. The third part Bib.3 contains those supporting books or papers that I refer to within the text.

The book is my view of the antieigenvalue theory and applications, presented chronologically from my creation of the theory to the present, and based chiefly upon my own ideas and research. The goal is to take the subject out to a wider audience.

I am indebted throughout my career to the technical typing of Elizabeth Stimmel and this book is no exception.

1.7   Exercises

1. Derive Schwarz's inequality: $|\langle u, v \rangle| \leq \|u\| \|v\|$.
2. Given an $n \times n$ SPD matrix A, derive its Euler equation $Ax_k = \lambda_k x_k$ from its Rayleigh quotient variational characterization of its eigenvalues

$$\lambda_k = \min_{\substack{x \neq 0 \\ x \perp x_1,\ldots,x_{k-1}}} \frac{\langle Ax, x \rangle}{\langle x, x \rangle}.$$

3. Show that every $n \times n$ matrix A has a spectral representation

$$A = \sum_{i=1}^{n} c_i P_i$$

   where the P_i are (generally oblique) projections: $P_i^2 = P_i$.
4. Review the definitions and main properties of (a) bounded operators A, and (b) closed unbounded operators A, in both (i) Banach spaces, and (ii) Hilbert spaces.
5. Show that for all matrix products, the spectra $\sigma(AB)$ and $\sigma(BA)$ are the same except possibly with the addition of the complex point 0.
6. Do some reading on computational linear algebra, matrix statistics, quantum mechanics, and financial instruments.

Chapter 2

The Original Motivation: Operator Semigroups

Perspective

Eventually, to really understand any subject, one needs to have gradually absorbed its history. Usually these days, none of us have the time to do that, for any subject. However, for the antieigenvalue theory, I know its history intimately, having originated the theory and stayed with it. You could even call my continued interest in it a labor of love: in that few paid attention to it for its first twenty years, and that for its last twenty years, it took a sustained effort on my part to make its beauty and usefulness evident through a range of applications to other parts of science.

Such philosophizing may be all well and good, but there is a much better imperative for presenting the originating application in this Chapter 2. It was that application to multiplicative perturbation of infinitesimal generators of contraction semigroup solutions for abstract initial value problems that led me to the key inequality (∗) which I exposed in Chapter 1. As revealed there, the key inequality (∗) led to the essential ingredients of my antieigenvalue theory and later development of the full operator trigonometry.

In retrospect, I was a young mathematician pushing for a multiplicative perturbation theory to accompany an existing additive perturbation theory. Who could have predicted that from such an abstruse context would emerge a new chapter in linear algebra?

2.1   Abstract Initial Value Problems

If you consider for example the heat equation initial value problem (IVP)

$$
\begin{cases}
\dfrac{\partial u}{\partial t} = \Delta u, & t > 0, \\[2mm]
u(x,0) = u_0(x) & \text{given,}
\end{cases}
$$

where $u = u(x,t)$ is the temperature of some physical body and where Δu denotes the Laplacian operator $\partial^2/\partial x_1^2 + \cdots + \partial^2/\partial x_n^2$, then in one space dimension (e.g., an infinite cooling rod) the classical solution for $-\infty < x < \infty$ and all forward time t is given by

$$
u(x,t) = \int_{-\infty}^{\infty} \frac{e^{-(x-y)^2/4t}}{\sqrt{4\pi t}} u_0(y)\,dy.
$$

Thus exponentiation has entered in a natural, explicit, and concrete way. Moreover in this example we have the exact Green's function kernel in the integral.

As is well known, the heat equation is one of the most important objects in mathematics. It figured as a fundamental concept in the recently completed proof of the topological Poincaré Conjecture. The Black–Scholes partial differential equation which so revolutionized high finance during the last forty years owes much of its implementational importance to its transformation by change of variable to the heat equation and then on to the Gaussian normal probability distribution from which option prices are calculated.

The abstract theory of such initial value problems is sometimes called Hille–Yosida theory after two of its principal pioneers. It may be regarded as a branch of functional analysis which grew out of a need to place a number of such partial differential equation problems into a rigorous and general mathematical setting. For more information and background, see the books Hille and Phillips (1957) and Kato (1976). Another classic reference is the book by Yosida (1968) or one of its later editions. For more about the heat equation and the Black–Scholes equation see for example the book Gustafson (1999f).

2.2   The Hille–Yosida–Phillips–Lumer Theorem

The version due to Lumer and Phillips (1961) of the Hille–Yosida Theorem characterizing the infinitesimal generators A of contraction semigroup solutions $u(t) = Z_t u_0$ for initial value problems such as the heat equation or the Schrodinger equations of quantum mechanics, is as follows.

Theorem 2.1 (Infinitesimal Generators). *Let A be a densely defined linear operator with domain $D(A)$ and range $R(A)$ in a Banach space X. Then A generates a continuous contraction semigroup Z_t on X if and only if the real part of its numerical range $\mathrm{Re}[Ax, x] \overset{<}{=} 0$ for all x in $D(A)$ and the range $R(I - A) = X$.*

Here $[y, x]$ denotes a semi-inner product structure on the Banach space. We may just say that such A be called dissipative operators (opposite of accretive operators). Thus in particular, the spectrum $\sigma(A)$ must lie in closed left half-complex plane. It follows that the operator $\lambda I - A$ is invertible and onto X for all λ with real part $\mathrm{Re}\,\lambda > 0$.

We let $G(1, 0)$ denote all such infinitesimal generators relevant to that particular Banach space.

2.3   The Rellich–Kato–Nelson–Gustafson Theorem

The most important Additive Perturbation Theorem is the following, due to Rellich and Kato in Hilbert space and Nelson and Gustafson in Banach space.

Theorem 2.2 (Additive Perturbation). *Let A be a contraction semigroup generator. Consider regular perturbation B, i.e., with $D(B) \supset D(A)$, which is relatively small with respect to A: there exist constants a and b, with $b < 1$, such that*

$$\|Bx\| \overset{<}{=} a\|x\| + b\|Ax\|$$

for all x in $D(A)$. Then $A + B$ remains a contraction semigroup generator if and only if $A + B$ is dissipative.

This theorem is extremely important in quantum mechanics, for a version of it due independently by Kato and Rellich in the 1940s established the selfadjointness of all of the basic Schrodinger Hamiltonian operators of chemical physics. For example, if one considers the Hydrogen operator

$$Hu = -\frac{1}{2}\Delta u - \frac{Z}{r}u,$$

then the Laplacian term is A and the Coulomb potential term is B. You may see in my book Gustafson (1999f) how use of the Schwarz and Sobolev inequalities establishes the selfadjointness of H. Also there, going further, one may use the same inequalities to show that the atom cannnot collapse. This fact is sometimes referred to as the stability of matter. See Kato (1976) for more details, citations to his and Rellich's original work, and further implications of it.

Nelson (1964) extended the Rellich–Kato Theorem to Banach space and relatively small additive perturbations B with bound $b < 1/2$. I extended it to $b < 1$ in my first published mathematics paper Gustafson (1966).

2.4   The Multiplicative Perturbation Theorem

Following a result by Dorroh (1966) in which he successfully multiplicatively perturbed a contraction semigroup generator A in a particular Banach space, I wanted a general result and obtained the following in [Gustafson (1968a)].

Theorem 2.3 (Multiplicative Perturbation). *Let A be a contraction semigroup generator. Suppose that B is a bounded operator such that for some $\epsilon > 0$, $\|\epsilon B - I\| < 1$. Then the multiplicatively perturbed operator BA is a contraction semigroup generator if and only if it is dissipative.*

There are some applications of my multiplicative perturbation result but it is not nearly as widely employed as the additive perturbation theory. However in this chapter the point is to show how it led to the creation of the antieigenvalue theory.

That may be seen by how I proved the Multiplicative Perturbation Theorem. I resorted to the Additive Perturbation Theorem. Inserting a free parameter $\epsilon > 0$, which will not alter whether or not ϵBA is a generator, one writes $\epsilon BA = A + (\epsilon B - I)A$ and one notes that $\|(\epsilon B - I)Ax\| \overset{\leq}{=} \|\epsilon B - I\|\|Ax\|$ implies by the Additive Perturbation Theorem that $(\epsilon B - I)A$ is an acceptable additive perturbation provided that $b \equiv \|\epsilon B - I\| < 1$ for some $\epsilon > 0$. Indeed one finds that the norm curve $\|\epsilon B - I\|$ dips below 1 for an interval $(0, \epsilon_c)$ whenever B is strongly accretive. Applying the same additive decomposition to the multiplicative perturbation in the semi-inner product, one finds

$$\text{Re}\,[\epsilon B A x, x] = \text{Re}\,[(\epsilon B - I)A x, x] + \text{Re}\,[A x, x]$$
$$\stackrel{\leq}{=} \|\epsilon B - I\|\,\|A x\|\,\|x\| + \text{Re}\,[A x, x]$$

from which one obtains the sufficient condition for ϵBA, and hence BA, to be dissipative:

$$\min_{\epsilon > 0} \|\epsilon B - I\| \stackrel{\leq}{=} \inf_{\substack{x \in D(A) \\ x \neq 0}} \text{Re}\, \frac{[(-A)x, x]}{\|A x\|\,\|x\|}.$$

This is, of course, the key inequality (∗) which led me to the antieigenvalue operator trigonometry.

As I mentioned in Sec. 1.3, I was not happy with the restriction to bounded B in the Multiplicative Perturbation Theorem. Accordingly, Lumer and I extended it to unbounded B in [Gustafson (1972b)]. Also Calvert and I extended it to nonlinear semigroup theory in [Gustafson (1972c)]. Sato and I developed mixed additive–multiplicative versions for Markov processes in [Gustafson (1969c)]. However, I put no operator trigonometry into those works. For further details of this early semigroup perturbation work and results obtained therein, see the abstracts in the AMS Notices [Gustafson (1967)] and the papers [Gustafson (1968a,b,c; 1969a)].

Because the left-hand side of the key inequality (∗) had essentially reduced our context to bounded B, I wondered about the right-hand side of the inequality. For that, in [Gustafson (1969b)], Zwahlen and I proved the disappointing result that the cosine of unbounded accretive operators A in a Hilbert space is zero. This result is quite interesting in itself as a new trigonometric distinction between bounded and unbounded accretive operators in a Hilbert space. And it does not always hold in general Banach spaces: I give examples in some of my papers. But in this early history of the antieigenvalue theory, these limitations to bounded B and A led me away from the semigroup perturbation theory and increased my attention toward the antieigenvalue operator trigonometry as a subject of interest in itself.

The objects were now bounded strongly accretive operators — the generalization of A and B both SPD matrices. The immediate question before me was: "When are such products BA accretive?" I did not find in the literature at that time even a natural answer for A and B both SPD matrices.

2.5 When are Positive Operator Products Positive?

I must confess that in all the semigroup perturbation work in those early days, I kept my antieigenvalue trigonometric ideas and results and evolving operator trigonometry to myself. Besides, my collaborators did not seem to be interested in it. A counterexample to those statements concerns my young colleague Charlie McCarthy at the University of Minnesota when I was there in 1966–1968. I am indebted to Charlie who in that period was very enthusiastic when I shared with him the elements of my beginning operator trigonometry, and especially about my Min-Max Theorem. It was Charlie who insisted I protect priority through my AMS Notices [Gustafson (1967, 1968d)]. On my side I was already heading off to quantum scattering theory, e.g., see the paper [Gustafson (1973a)] with Peter Rejto, another colleague at the University of Minnesota with whom I interacted. I have never thanked Charlie, and to my memory never saw him again after I left Minnesota. So let me acknowledge his optimism and energy here. McCarthy at that time was publishing his very nice Cp inequalities paper [McCarthy (1967)].

After this aside, let us turn then to the issue of positivity of matrix products BA. Let us illustrate with an example what the quantitative improvement the operator-trigonometric sufficient condition

$$\sin \phi(B) \stackrel{\leq}{=} \cos \phi(A)$$

can make. For simplicity, let A and B be SPD matrices where A's spectrum has smallest eigenvalue $\lambda_1 = 1/2$ and largest eigenvalue $\lambda_n = 1$. Let B's largest eigenvalue also be $\lambda_n = 1$. What can B's smallest eigenvalue λ_1 be such that Re BA will still be nonnegative?

Before my antieigenvalue theory, one could easily see from the condition number criteria

$$\|I - B\| = 1 - \lambda_1^B \stackrel{\leq}{=} \frac{\lambda_1^A}{\lambda_n^A} = \frac{1}{2}$$

that λ_1^B can go down to $1/2$. If we now sharpen that criteria by putting in $\sin \phi(B)$, we get

$$\sin \phi(B) = \|I - \epsilon_m B\| = \frac{1 - \lambda_1^B}{1 + \lambda_1^B} \stackrel{\leq}{=} \frac{1}{2}$$

and see that λ_1^B can go down to $1/3$. If we further sharpen by replacing the right-hand side by $\cos \phi(A)$, we arrive at the completely trigonometric criteria

$$\sin \phi(B) = \frac{1 - \lambda_1^B}{1 + \lambda_1^B} \stackrel{\leq}{=} \cos \phi(A) = \frac{2\sqrt{2}}{3} = 0.9428$$

from which λ_1^B can go down to 0.0295, approximately. So we can assert that roughly 97% of the B are okay!

After I published my trigonometric criteria for positive operator products, Gilbert Strang sent me his paper [Strang (1962)]. His generous comment at the time was that my approach was sharper. Many years later I returned to positive operator products in [Gustafson (2004c)] where I study the more general concept of interaction antieigenvalue which I had casually mentioned over thirty years earlier in [Gustafson (1972a)]. In [Gustafson (2004c)] I also compare the relative sharpnesses of my positive operator product bounds, Strang's bounds, the Greub–Rheinboldt bounds, and J. Stampfli's operator derivation bounds [Stampfli (1970)]. But there are no trigonometric considerations in the papers by those other authors. And, generally speaking, my trigonometric criteria are sharper.

2.6 Nonnegative Contraction Semigroups

Markov processes and more general stochastic processes possess a different nonnegativity property: that of positivity in the Banach lattice sense. In joint work [Gustafson (1969c)] with Ken-ito Sato while he was visiting the University of Minnesota, we extended my multiplicative perturbation theory to the Banach lattice setting. However as I have already indicated earlier in this book, my operator trigonometry seems to live naturally in Hilbert space, and is less applicable in Banach spaces. This state of affairs should not be surprising since usual spectral theory is much richer in Hilbert spaces than it is in Banach spaces. Such is even more the case for my antieigenvalue spectral theory due to its very nature of being a spectral theory of angles. Hilbert space is rich in angular content and orthogonal bases and unitary and symmetric linear transformations. The same cannot be said for Banach spaces.

The story of the joint paper Gustafson (1969c) is this. Sato gave a colloquium lecture about the Banach lattice nonnegative contraction semigroup generators as they occur in Markov process theory. There, the Lumer–Phillips necessary condition of dissipativeness is replaced with another tangent-functional condition called dispersiveness. Immediately during his lecture, I saw that if one could prove a lemma: a 0-dispersive wide sense operator B is dissipative in at least one semi-inner product on the Banach space, then in a few lines one could generalize my multiplicative perturbation results to Banach lattices. But we could not prove my conjectured lemma for general Banach lattices. Therefore we had to obtain the analogous perturbation results the hard way, reproving the theorems in the Banach lattice structures.

Although my conjecture about dispersiveness implying dissipativeness has been shown to be true for many Banach lattices, I have not followed the literature and do not know if it has been established in general, or if a counterexample has been produced anywhere.

In similar vein, I have no idea as to whether anyone has tried to imitate my operator trigonometry within the dispersive structures of Banach lattices rather than in the dissipative structures of Banach spaces.

Commentary

The naturally encountered restrictions to bounded perturbation B and then to bounded generators A and in a Hilbert space for the use of my evolving antieigenvalue operator trigonometry for the semigroup perturbation theory was definitely disappointing to me. But as the more interesting applications to be presented later in this book will make clear, the most interesting uses of my trigonometry concern symmetric matrices or Hermitian operators on a Hilbert space. Thus the principal merit of the original problem in semigroup generator perturbation as concerns the antieigenvalue theory is that it produced the two key entities in the inequality (∗) which became $\cos \phi(A)$ and $\sin \phi(A)$ for use later in Hilbert space applications. That is, it gave birth to the antieigenvalue theory and the operator trigonometry.

2.7   Exercises

1. Show that the Laplacian operator $Au = \Delta u$ is dissipative in $L^2(\mathbb{R})$ if you consider u's which vanish at infinity or on the boundary of the cooling body Ω. It is strongly dissipative on bounded domains of Ω.

2. Review for yourself from an elementary ordinary differential equations book the ways in which one may exponentiate the system

$$
\begin{bmatrix} \dot{x_1}(t) \\ x_2(t) \\ \vdots \\ x_n(t) \end{bmatrix} = \begin{bmatrix} a_{11} & \dots & a_{1n} \\ a_{21} & \dots & a_{2n} \\ & \ddots & \\ a_{n1} & \dots & a_{nn} \end{bmatrix} \begin{bmatrix} x_1(t) \\ x_2(t) \\ \vdots \\ x_n(t) \end{bmatrix}
$$

to the semigroup solution $x(t) = e^{tA}x(0)$.

3. Read about the Black–Scholes partial differential equation of modern derivatives finance, and how it may be reduced to the heat equation and then to the Gaussian (normal) probability distribution to price call and put options.

4. Use the Rellich–Kato Additive Perturbation Theorem to establish the selfadjointness of the Hydrogen atom Hamiltonian, as mentioned in Sec. 2.2 above.

5. Prove that if A (bounded or unbounded) is a positive selfadjoint operator and B is an accretive bounded operator with which A commutes, then BA is accretive. In addition, suppose that moreover B is selfadjoint and strongly positive. Then $BA = AB$ and is selfadjoint and strongly positive.

6. One can make a closed unbounded operator into a bounded one by using the operator graph norm. Prove the following interesting fact for possible use in our contraction semigroup context to treat unbounded operators there.

Lemma 2.1. *Let A be a closed densely defined operator in a Hilbert space $\mathcal{H}$ which is the infinitesimal generator of a contraction semigroup $Z_t = e^{tA}$. Let $\mathcal{H}_A$ be the Hilbert space $D(A)$ equipped with the inner product*

$$
\langle x, y \rangle_A = \langle x, y \rangle + \langle Ax, Ay \rangle.
$$

Then Z_t remains a contraction semigroup on $\mathcal{H}_A$.

Chapter 3

The Essentials of Antieigenvalue Theory

Perspective

Chapter 1 already introduced some of the essential entities of the antieigen-value theory. In this chapter we wish to develop those and other essentials of the antieigenvalue theory in a more systematic and rigorous way. A chronological style of exposition continues to be both natural and efficient.

In Chapters 1 and 2 we described how the original application brought forth the key inequality (∗) which then led to my development of the antieigenvalue concepts and the operator trigonometry. Let us repeat here that key inequality, in both its full and specialized forms

$$\inf_{\substack{\infty > \epsilon > -\infty \\ \|x\| \stackrel{<}{=} 1}} \sup \|(\epsilon B - I)x\| \stackrel{\leq}{=} \inf_{x \neq 0} \frac{\mathrm{Re}\,\langle Ax, x \rangle}{\|Ax\| \|x\|}, \tag{∗1}$$

$$\min_{\epsilon > 0} \|\epsilon B - I\| \stackrel{\leq}{=} \min_{x \neq 0} \frac{\langle Ax, x \rangle}{\|Ax\| \|x\|}. \tag{∗2}$$

The (∗1) version concerns accretive operators A and B in a Banach space, if we understand the $\langle y, x \rangle$ symbol loosely to mean any semi-inner product you may be working with on that Banach space. We will not overly concern ourselves with the technicalities of semi-inner product structures in this book. If you want to know more, we recommend the papers [Lumer and Phillips (1961)] and my later paper with J. P. Antoine [Gustafson (1981a)]. The (∗2) version can be thought of as for finite-dimensional SPD matrices A and B.

It is evident that the left-hand side of (*) has to do with properties of norm curves in Banach and Hilbert spaces. It is also evident that the right-hand side of (*) is a variational quantity. The short statement of my antieigenvalue theory is that it extends the variational Rayleigh–Ritz theory of eigenvalues and eigenvectors to also include antieigenvalues and antieigenvectors and then connects that extended spectral theory to the geometry of normed linear spaces.

3.1 Convexity Properties of Norm Geometry

If you look at my earlier papers, especially [Gustafson (1968c,d)], you will see that originally the norm curve $\|\epsilon B - I\|$ I called $g(\epsilon, B)$ and its minimum I called $g_m(B)$. That was because from the viewpoint of the geometry of Banach spaces, those entities are tangent functionals. I do not want to embark here into a journey into Banach space geometry, but at that time I was quite familiar with it. Beyond the references I gave in [Gustafson (1968c)], I also recommend the paper [Cudia (1964)] and the further citations therein.

What I needed in my study of the $\|\epsilon B - I\|$ norm curve was that it is continuous and convex in ϵ. Moreover at every ϵ both the left and right derivatives exist and are equal except for a countable set of ϵ. From these well-known properties, we see that there can be three kinds of minima $g_m(B)$ that can occur. If $g_m(B)$ occurs at a differentiable point of the curve, then we have either a flat or a cup shape there. It is easy to show that for a strongly accretive B on a Hilbert space, there can be no interval flat minimum. We will show this in the next section. Simple examples show that normal operators may have a cup-shaped minimum. The most interesting and useful shape of minimum is that of a V-like corner, as in Fig. 1.1 in the first chapter. It goes without saying that all of this discussion is for B not a multiple of the identity I. So in Hilbert space our strongly accretive operators have norm curves $\|\epsilon B - I\|$ which all decrease from the value 1 at $\epsilon = 0$ to some unique minimum $g_m(B) = \|\epsilon_m B - I\|$ whose value is strictly between 1 and 0 and then increase to the right of ϵ_m to (relatively soon) again attain the value 1 and increase to the right thereafter. They also increase to the left of $\epsilon = 0$ as ϵ becomes more negative. But there, their values exceed 1.

3.2   The Min-Max Theorem

I proved Theorem 1.1 in 1968 and announced it in [Gustafson (1968d)] just because I wanted the left-hand side of the inequality (∗) to be trigonometric so that the multiplicative perturbation BA could be guaranteed to be accretive by the inequality (∗) now seen trigonometrically:

$$\sin \phi(B) \overset{\leq}{=} \cos \phi(A).$$

Then at the 1969 *Third Symposium on Inequalities* in Los Angeles, I presented my proof, which eventually appeared in the paper [Gustafson (1972a)]. When I returned to the antieigenvalue theory in the early 1990s, in the paper [Gustafson (1995a)] dedicated to my colleague John Maybee upon his retirement, I returned to my Min-Max Theorem and gave the proof again there. Not only did I think that the fundamental result was a nice one to put into a paper dedicated to a very close friend (who unfortunately died four years later), but I must admit that at that time, I also was refreshing my mind as I was returning to the antieigenvalue theory.

Then of course the proof of the Min-Max Theorem is given again in full detail in the two 1997 books [Gustafson (1997a,d)]. What should we do here?

Because I view this Min-Max Theorem as the crux of my operator trigonometry, in the following I will further elaborate two of its key points. We recall the theorem as stated earlier in Sec. 1.3: if A is a strongly accretive bounded operator on a Hilbert space, then

$$\sup_{\|x\| \overset{\leq}{=} 1} \inf_{-\infty < \epsilon < \infty} \|(\epsilon A - I)x\|^2 = \inf_{\epsilon > 0} \sup_{\|x\| \overset{\leq}{=} 1} \|(\epsilon A - I)x\|^2. \qquad (\ast\ast)$$

First let us recall its general proof. For the reader's convenience I have depicted the proof in Fig. 3.1. A is a strongly accretive operator on a Hilbert space and the convex curve $\|\epsilon A - I\|$ has left and right derivatives for almost all $\epsilon > 0$, so its unique minimum $\|\epsilon A - I\|$ is attained at either a lower corner or at the bottom of a cup. We only consider the corner case here: the cup case is easier and less informative. Although we may not know exact minimizing vectors, we do know that for arbitrary small $\delta > 0$ we can always find norm-one vectors x_1 and x_2 such that to the left and right of ϵ_m, respectively, $\|(\epsilon A - I)x_1\|^2$ and $\|(\epsilon A - I)x_2\|^2$ approximate to within δ the curve $\|\epsilon A - I\|^2$ near ϵ_m. These are the curves $\|(\epsilon B - I)x_\ell\|^2$ and

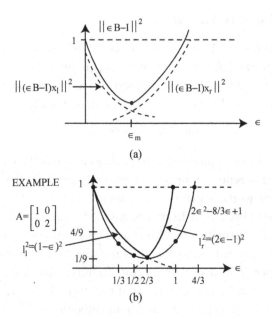

Fig. 3.1. Sketch of proof of Min-Max Theoren.
(a) The left and right norm-approximating curves.
(b) The resulting antieigenvector curve for Example 1.1.

$\|(\epsilon B - I)x_r\|^2$ in Fig. 3.1(a) for a general strongly accretive matrix B, where I used the notation x_{left} and x_{right} for clarity. For the matrix A of Example 1.1 as shown in Fig. 3.1(b), they are exactly the curves $(1 - \epsilon)^2$ and $(2\epsilon - 1)^2$, respectively. Therefore we take $x = \xi x_1 + \eta x_2$ where ξ and η are real,

$$1 = \|x\|^2 = \xi^2 + \eta^2 + 2\eta\xi \, \mathrm{Re} \, \langle x_1, x_2 \rangle$$

and

$$\|(\epsilon_m A - I)x\|^2 \gtrless \|\epsilon_m A - I\|^2 - \delta + 2\xi\eta C$$

where

$$C = \mathrm{Re} \{ \langle (\epsilon_m A - I)x_1, (\epsilon_m A - I)x_2 \rangle - (\|\epsilon_m A - I\|^2 - \delta)\langle x_1, x_2 \rangle \}.$$

By choosing ξ and η in the appropriate quadrant we can assure that $2\xi\eta C \gtrless 0$. Because $\|(\epsilon_m A - I)x\|^2$ must always lie below $\|(\epsilon_m A - I)\|^2$ yet also above the just obtained lower bound we have obtained an approximate minimizing vector x with $\|(\epsilon_m A - I)x\|^2$ within arbitrarily small $\delta > 0$ just below $\|\epsilon_m \mathbf{A} - \mathbf{I}\|^2$. Since the Min-Max equality states that

$$\sup_{\|\mathbf{x}\| \leq 1} \inf_{\epsilon} \|(\epsilon \mathbf{A} - \mathbf{I})\mathbf{x}\|^2 = \inf_{\epsilon > 0} \sup_{\|\mathbf{x}\| \leq 1} \|(\epsilon \mathbf{A} - \mathbf{I})\mathbf{x}\|^2,$$

we have established that the left-hand side attains the right-hand side. That establishes the Min-Max Theorem because the right-hand side is *a priori* never less than the left-hand side.

Moreover one then checks that $\epsilon_m(\mathbf{x})$ can be made arbitrarily close to $\epsilon_m(\mathbf{A})$, as it must be if the pair $(\epsilon_m(\mathbf{x}), \|(\epsilon \mathbf{A} - \mathbf{I})\mathbf{x}\|^2)$ is to converge to $(\epsilon_m(\mathbf{A}), \|(\epsilon \mathbf{A} - \mathbf{I})\|^2)$. My proof of that reveals, if I might say so, a rather nice use of elementary ellipses and hyperbolas. Then to go further and ask that $\epsilon_m(\mathbf{x})$ can be made to exactly equal $\epsilon_m(\mathbf{A})$, I show such to be equivalent to the relation

$$\xi^2 \left\{ \operatorname{Re} \langle \mathbf{A}\mathbf{x}_1, \mathbf{x}_1 \rangle \left(1 - \frac{\epsilon_m}{\epsilon_1} \right) \right\} + \eta^2 \left\{ \operatorname{Re} \langle \mathbf{A}\mathbf{x}_2, \mathbf{x}_2 \rangle \left(1 - \frac{\epsilon_m}{\epsilon_1} \right) \right\}$$
$$- 2\xi\eta \operatorname{Re} \langle \mathbf{A}\mathbf{x}_1, (\epsilon_m \mathbf{A} - \mathbf{I})\mathbf{x}_2 \rangle = 0.$$

Here $\epsilon_1 = \operatorname{Re} \langle \mathbf{A}\mathbf{x}_1, \mathbf{x}_1 \rangle / \|\mathbf{A}\mathbf{x}_1\|^2$ is where the curve $\|(\epsilon \mathbf{A} - \mathbf{I})\mathbf{x}_1\|^2$ attains its minimum $1 - (\operatorname{Re} \langle \mathbf{A}\mathbf{x}_1, \mathbf{x}_1 \rangle / \|\mathbf{A} - \mathbf{I}\|)^2$; likewise ϵ_2 is where $\|(\epsilon \mathbf{A} - \mathbf{I})\mathbf{x}_2\|^2$ achieves its minimum.

Now I want to specialize here to the situation wherein $\mathbf{A}$ is an $n \times n$ SPD matrix. Then we know just from the norm curve $\|\epsilon \mathbf{A} - \mathbf{I}\|$ that $\mathbf{x}_1$ and $\mathbf{x}_2$ in the Min-Max Theorem proof may be taken to be any norm-one eigenvectors $\mathbf{x}_1$ and $\mathbf{x}_n$ corresponding to the smallest and largest eigenvalues λ_1 and λ_n, respectively. See the example given in Fig. 1.1. Then $\epsilon_1 = 1/\lambda_1$, $\epsilon_2 = 1/\lambda_n$ and $\epsilon_m(\mathbf{A}) = 2/(\lambda_1 + \lambda_n)$, and the expression above simplifies to

$$\xi^2 \left\{ \lambda_1 \left(1 - \frac{2\lambda_1}{\lambda_1 + \lambda_n} \right) \right\} + \eta^2 \left\{ \lambda_n \left(1 - \frac{2\lambda_n}{\lambda_1 + \lambda_n} \right) \right\} = 0,$$

or seen more trigonometrically, to the expression

$$\sin \phi(\mathbf{A})[\lambda_1 \xi^2 - \lambda_n \eta^2] = 0.$$

Also the expression $\|(\epsilon_m \mathbf{A} - \mathbf{I})\mathbf{x}\|^2$ simplifies to

$$\|(\epsilon_m \mathbf{A} - \mathbf{I})\mathbf{x}\|^2 = \sin^2 \phi(\mathbf{A})[\xi^2 + \eta^2].$$

We have of course used the orthogonality $\langle \mathbf{x}_1, \mathbf{x}_n \rangle = 0$. Thus because $\xi^2 + \eta^2 = 1$ we see that the minimum attaining vector $\mathbf{x} = \xi\mathbf{x}_1 + \eta\mathbf{x}_n$ may be found from the conic system

$$\begin{cases} \xi^2 + \eta^2 = 1, \\ \lambda_1 \xi^2 - \lambda_n \eta^2 = 0. \end{cases}$$

Solving that yields

$$\xi = \pm \left(\frac{\lambda_n}{\lambda_1 + \lambda_n} \right)^{1/2}, \quad \eta = \pm \left(\frac{\lambda_1}{\lambda_1 + \lambda_n} \right)^{1/2}.$$

These are the correct coefficients of the antieigenvectors $x_\pm$.

Looking again at Fig. 3.1(b) for the matrix A of Example 1.1, because the left and right norm-approximating curves $(1 - \epsilon)^2$ and $(2\epsilon - 1)^2$ are exactly the norm curve $\|\epsilon A - I\|^2$, we see that we did not even need the more general convexity-based approximation proof above. Plugging in the antieigenvector coefficients ξ and η found above to the antieigenvector $x_+ = \xi x_1 + \eta x_2$ we find the true antieigenvector parabola for A to be

$$\|(\epsilon A - I)x_+\|^2 = \frac{2}{3}(1 - \epsilon)^2 + \frac{1}{3}(2\epsilon - 1)^2$$

$$= 2\epsilon^2 - \frac{8}{3}\epsilon + 1$$

as shown there.

Because I have encountered over the years a certain amount of wonder and in some instances even doubt as to my proof of my Min-Max Theorem, let me make a few comments here. First, my use of, essentially, least squares convexity arguments using what I would call Hilbert space parabolas, was just the way in which I saw that I could prove the theorem. Remember that my original goal was to make $\min_\epsilon \|B - I\|$ trigonometric. Then I wanted to get the most turned vectors, which I called the antieigenvectors. My proof is constructive and gets them, either exactly or approximately. Thus for $n \times n$ SPD matrices, the first key insight I want to bring out now, already evident in the above proof of the Min-Max Theorem, is that the proof itself identifies the minimizing vectors to be the linear combination of the first eigenvector weighted by the last eigenvalue plus the last eigenvector weighted by the first eigenvalue. You can choose whichever convention is your habit for ordering the eigenvalues: but be sure to counterweight the eigenvectors to get the antieigenvectors!

Thus we have proved the assertion made earlier in Sec. 1.3 that the most turned vectors, the antieigenvectors, are

$$x_{\pm} = \pm \left(\frac{\lambda_1}{\lambda_1 + \lambda_n} \right)^{1/2} x_n + \left(\frac{\lambda_n}{\lambda_1 + \lambda_n} \right)^{1/2} x_1.$$

Just think of them as the two extreme eigenvectors, weighted oppositely by their eigenvalues in order to achieve a "most twisted" angle when operated upon by the matrix A. The $(\lambda_1 + \lambda_n)$ denominators in the weights are just in there to make $x_{\pm}$ of norm one, given that both of the eigenvectors are of norm one. The eigenvectors can be chosen arbitrarily from their (possibly higher multiplicity) respective eigenspaces, but I caution you that they must both be taken of same norm. That in higher multiplicity eigenspaces you may choose any norm-one eigenvector was clarified in [Gustafson (2000d)]. Clearly you can put the $\pm$ on either the first or second component of the antieigenvector. Scalar multiples of $x_{\pm}$ clearly also turn maximally. But we may not linearly combine them to claim some kind of "antieigenspace."

The second key insight I wish to bring out is the following. First note that in the Min-Max equality ($**$), the right-hand side is $\|\epsilon A - I\|^2$. It is easy to check that in ($**$) all left-hand sides are bounded above by the right-hand side. Hence one already had a Min-Max inequality. Thus the task in the proof of the Min-Max Theorem can now be viewed as that of finding some vectors x which can bring the left-hand side up to the right-hand side.

But there is more to it. Let us just focus on the right-hand side of ($**$). Which are the vectors x which attain the right-hand side? It turns out that there are many. I state this result as I did recently in [Gustafson (2010d)]. I shall elaborate some of its further implications in Chapter 9.

Theorem 3.1 (General Optimum Theorem). *Let $x = c_1 x_1 + c_n x_n$ where x_1 is any norm-one eigenvector corresponding to λ_1 and x_n is any norm-one eigenvector corresponding to λ_n such that $|c_1|^2 + |c_n|^2 = 1$. Then*

$$\sin \phi(A) = \|(\epsilon_m A - I)x\|, \quad \epsilon_m = \frac{2}{(\lambda_1 + \lambda_n)}.$$

In contrast,

$$\cos \phi(A) = \min_{x \neq 0} \frac{\langle Ax, x \rangle}{\|Ax\| \|x\|}$$

is attained only for the special weights $|c_1|^2 = \lambda_n/(\lambda_1 + \lambda_n)$ and $|c_n|^2 = \lambda_1/(\lambda_1 + \lambda_n)$, i.e., only for the antieigenvectors $x_{\pm}$.

Proof. The proof may be obtained from a careful examination of the vectors x of norm one which "work" in the following two important relations:

$$\max_{\|x\|=1} \min_{\epsilon>0} \|(\epsilon A - I)x\|^2 = \max_{\|x\|=1} \left[1 - \left(\frac{\text{Re}\langle Ax, x \rangle}{\|Ax\|} \right)^2 \right]$$

$$= 1 - \min_{\|x\|=1} \left(\frac{\text{Re}\langle Ax, x \rangle}{\|Ax\|} \right)^2$$

$$= 1 - \cos^2 \phi(A),$$

and

$$\min_{\epsilon>0} \max_{\|x\|=1} \|(\epsilon A - I)x\|^2 = \min_{\epsilon>0} \|\epsilon A - I\|^2$$

$$= \sin^2 \phi(A).$$

The details may be checked by the reader.

The theorem (General Optimum Theorem) highlights how much more generally the $\sin \phi(A)$ functional is attained, as compared to the $\cos \phi(A)$ functional. One only needs to be on the unit ball in the combined first and last eigenspaces. Then I view the obtaining of the $\cos \phi(A)$ as a special selection from within that general two-component unit ball.

A final word about that. As we shall see later in this book, and as will be made especially evident in Chapter 6, when we compare solutions to the Euler equation with those to what I call the "inefficiency equation", one must take care in identifying which vectors optimize an expression. The above proof shows that the antieigenvectors do indeed maximize the left-hand side of the Min-Max (∗∗) equality. But that is not how I defined antieigenvectors. I originally defined them to minimize the right-hand side of the inequality (∗). For the $n \times n$ SPD matrix A case, recall from Sec. 1.3 that they were to be those vectors which achieved that Rayleigh-type variational quotient minimization

$$\min_{x \neq 0} \frac{\langle Ax, x \rangle}{\|Ax\| \|x\|}.$$

By Schwarz's inequality that quotient achieves its upper bound of one at eigenvectors. When the inequality (∗) fell into my lap from the semigroup perturbation problems, it was natural for me to see the above variational quotient minimum value as defining the largest angle through which A

could turn any vector. Accordingly, right at the inception of the antieigen-value theory I called that angle $\phi(A)$, and those most-turned vectors the antieigenvectors of A.

So the role of antieigenvectors that one should take within the context of the above General Optimum Theorem is simply that they are a few of the many which achieve $\sin \phi(A)$.

3.3   The Euler Equation

What then are the minimizers of my antieigenvalue variational quotient

$$\frac{\text{Re} \langle Ax, x \rangle}{\|Ax\| \|x\|},$$

and how does one obtain them? As I was seeing my infant antieigen-value theory as an extension of eigenvector thinking to new antieigen-vector thinking, naturally I thought of the above antieigenvalue quotient as analogous to the Rayleigh variational quotients defining eigenvalues:

$$\lambda_1 = \min_{x \neq 0} \frac{\langle Ax, x \rangle}{\langle x, x \rangle},$$

$$\lambda_k = \min_{\substack{x \neq 0 \\ x \perp x_1, \ldots, x_{k-1}}} \frac{\langle Ax, x \rangle}{\langle x, x \rangle},$$

where the x_k are the corresponding eigenvectors. As I reminded the reader in Exercise 1.2, to show that it is the eigenvectors which actually do the Rayleigh quotient minimizing, one needs to do a functional differentiation of the variational quotient and set that to zero to obtain the eigenvector Euler equation

$$Ax = \lambda x.$$

So I did the same thing in 1968–1969 for the antieigenvalue variational quotient. It was more complicated. I will give the derivation below. My resulting antieigenvalue Euler equation for the antieigenvectors turned out to be

$$2\|Ax\|^2 \|x\|^2 (\text{Re} A)x - \|x\|^2 \text{Re} \langle Ax, x \rangle A^* Ax - \|Ax\|^2 \text{Re} \langle Ax, x \rangle x = 0.$$

For A an $n \times n$ SPD matrix and $\|x\| = 1$, this becomes

$$\frac{A^2 x}{\langle A^2 x, x \rangle} - \frac{2Ax}{\langle Ax, x \rangle} + x = 0.$$

I did not like this nonlinear equation. I did present it in my lecture at the 1969 *Third Symposium on Inequalities.* But in the proceedings paper [Gustafson (1972a)] you will find me only saying "let us comment, however, that the Euler equations for the antieigenvalues are unfortunately nonlinear."

As I write these lines, it seems to me that I have never used the above antieigenvector Euler equation to find the antieigenvectors. It was easier to divine them for special cases for selfadjoint matrices A as in [Gustafson (1968b)], or see them as extractable from the Min-Max Theorem proof as I showed in the previous section. Nonetheless when I returned to the subject in the 1990s, I felt it incumbent upon me to write up my 1969 rough notes deriving the Euler equation, and publish them. This I did in [Gustafson (1994b)]. Also I put my Euler equation proof into the two 1997 books [Gustafson (1997a,d)].

Then when I applied my antieigenvalue theory to matrix statistics in 1999 [Gustafson (1999e; 2002a)] I found an interesting parallel between my Euler equation for antieigenvectors and an equation for the efficiency of a linear estimator in statistics. This will be further explained in Chapter 6. Moreover, I knew from the beginning that my Euler equation for antieigenvectors is also satisfied by eigenvectors. Because it constitutes a significant extension of the usual Rayleigh variational theory for eigenvectors to now also include the antieigenvectors, for completeness I will give the proof here.

A being any bounded strongly accretive operator on any Hilbert space, I wanted a full functional derivation, e.g., not just a finite-dimensional derivation as I gave in the answer to Exercise 1.2 to find the Euler equation for the eigenvectors of an $n \times n$ selfadjoint matrix A. Let

$$\mu(u) = \operatorname{Re} \frac{\langle Au, u \rangle}{\|Au\| \|u\|}.$$

To find the Euler equation we consider the quantity

$$\frac{d\mu}{dw}\Big|_{\epsilon=0} = \lim_{\epsilon \to 0} \frac{1}{\epsilon} \left(\frac{\operatorname{Re} \langle A(u + \epsilon w), u + \epsilon w \rangle}{\langle A(u + \epsilon w), A(u + \epsilon w) \rangle^{1/2} \langle u + \epsilon w, u + \epsilon w \rangle^{1/2}} \right.$$
$$\left. - \frac{\operatorname{Re} \langle Au, u \rangle}{\langle Au, Au \rangle^{1/2} \langle u, u \rangle^{1/2}} \right).$$

Let the expression on the right-hand side be denoted $R_A(u, w, \epsilon)$. We have

$$\epsilon R_A(u, w, \epsilon)$$

$$= [\text{Re} \langle Au, u \rangle + 2\epsilon \, \text{Re} \, \langle (\text{Re} \, A)u, w \rangle + \epsilon^2 \langle Aw, w \rangle]$$

$$\times \langle Au, Au \rangle^{1/2} \langle u, u \rangle^{1/2} \div D$$

$$- [\langle Au, Au \rangle + 2\epsilon \, \text{Re} \, \langle Au, Aw \rangle + \epsilon^2 \langle Aw, Aw \rangle]^{1/2}$$

$$\times [\langle u, u \rangle + 2\epsilon \, \text{Re} \, \langle u, w \rangle + \epsilon^2 \langle w, w \rangle]^{1/2} \, \text{Re} \langle Au, u \rangle \div D,$$

where D is the common denominator

$$D = \langle A(u + \epsilon w), A(u + \epsilon w) \rangle^{1/2}$$

$$\times \langle u + \epsilon w, u + \epsilon w \rangle^{1/2} \langle Au, Au \rangle^{1/2} \langle u, u \rangle^{1/2}.$$

At this point, in deriving the Euler equations for eigenvalues of a self-adjoint operator, one gets a fortuitous cancelation of the ϵ-independent terms in the expression analogous to $\epsilon R_A(u, w, \epsilon)$, and the Euler equation for eigenvectors immediately follows. Although that situation does not occur here, we may attempt to mimic it by expanding the two square root bracket expressions of the second numerator term,

$$|\langle Au, Au \rangle + x(\epsilon)|^{1/2} = \langle Au, Au \rangle^{1/2} + \frac{1}{2} \langle Au, Au \rangle^{-1/2} x(\epsilon)$$

$$- \frac{1}{8} \langle Au, Au \rangle^{-3/2} x^2(\epsilon) + \cdots$$

and

$$|\langle u, u \rangle + y(\epsilon)|^{1/2} = \langle u, u \rangle^{1/2} + \frac{1}{2} \langle u, u \rangle^{-1/2} y(\epsilon)$$

$$- \frac{1}{8} \langle u, u \rangle^{-3/2} y^2(\epsilon) + \cdots,$$

where $x(\epsilon)$ and $y(\epsilon)$ are the ϵ-dependent terms, respectively, and where ϵ is sufficiently small relative to $\langle Au, Au \rangle$ and $\langle u, u \rangle$, respectively. Then we obtain $\langle Au, Au \rangle^{1/2} \langle u, u \rangle^{1/2} \, \text{Re} \, \langle Au, u \rangle$ term cancelations, from which

$$\epsilon R_A(u, w, \epsilon) = [2\epsilon \, \text{Re} \, \langle (\text{Re} \, A)u, w \rangle + \epsilon^2 \langle Aw, w \rangle]$$

$$\times \langle Au, Au \rangle^{1/2} \langle u, u \rangle^{1/2} \div D$$

$$- \text{Re} \, \langle Au, u \rangle [\langle u, u \rangle^{1/2} r(\epsilon) + \langle Au, Au \rangle^{1/2} s(\epsilon)] \div D,$$

where $r(\epsilon)$ and $s(\epsilon)$ denote the remainder terms in the square root series expansions above, to be specific,

$$r(\epsilon) = \frac{1}{2}\langle Au, Au\rangle^{-1/2}x(\epsilon) - \frac{1}{8}\langle Au, Au\rangle^{-1/2}x^2(\epsilon) + \cdots,$$

$$s(\epsilon) = \frac{1}{2}\langle u, u\rangle^{-1/2}y(\epsilon) - \frac{1}{8}\langle u, u\rangle^{-3/2}y^2(\epsilon) + \cdots,$$

where

$$x(\epsilon) = 2\epsilon \operatorname{Re}\langle Au, Aw\rangle + \epsilon^2\langle Aw, Aw\rangle,$$

$$y(\epsilon) = 2\epsilon \operatorname{Re}\langle u, w\rangle + \epsilon^2\langle w, w\rangle.$$

We may now divide by ϵ, from which

$$D \cdot R_A(u, w, \epsilon) = [2\operatorname{Re}\langle(\operatorname{Re} A)u, w\rangle + \epsilon\langle Aw, w\rangle]\|Au\|\|u\|$$
$$- \operatorname{Re}\langle Au, u\rangle[\|Au\|^{-1}\|u\|\operatorname{Re}\langle Au, Aw\rangle + O(\epsilon)]$$
$$- \operatorname{Re}\langle Au, u\rangle[\|u\|^{-1}\|Au\|\operatorname{Re}\langle u, w\rangle + O(\epsilon)].$$

Note also that by the above expansions

$$D = [\|Au\| + O(\epsilon)][\|u\| + O(\epsilon)]\|Au\|\|u\| \to \|Au\|^2\|u\|^2 \quad \text{as } \epsilon \to 0.$$

Thus in the $\epsilon \to 0$ limit of $R_A(u, w, \epsilon)$, we arrive at

$$\left.\frac{d\mu}{dw}\right|_{\epsilon=0} = \frac{2\operatorname{Re}\langle(\operatorname{Re} A)u, w\rangle\|Au\|^2\|u\|^2}{\|Au\|^3\|u\|^3}$$
$$- \frac{\operatorname{Re}\langle Au, u\rangle[\|u\|^2\operatorname{Re}\langle Au, Aw\rangle + \|Au\|^2\operatorname{Re}\langle u, w\rangle]}{\|Au\|^3\|u\|^3}.$$

Setting this expression equal to zero yields

$$2\|Au\|^2\|u\|^2\operatorname{Re}\langle(\operatorname{Re} A)u, w\rangle$$
$$- \operatorname{Re}\langle Au, u\rangle[\|u\|^2\operatorname{Re}\langle A^*Au, w\rangle + \|Au\|^2\operatorname{Re}\langle u, w\rangle] = 0$$

for arbitrary w, and hence we have the Euler equation

$$2\|Au\|^2\|u\|^2(\operatorname{Re} A)u - \|u\|^2\operatorname{Re}\langle Au, u\rangle A^*Au - \|Au\|^2\operatorname{Re}\langle Au, u\rangle u = 0.$$

Theorem 3.2 (General Euler Equation). *The above Euler equation for vectors u which minimize the antieigenvalue functional $\mu(u)$ for strongly accretive bounded operators on a Hilbert space has the following additional properties. Scalar multiples of solutions are solutions. When A is selfadjoint or normal, the Euler* *equation is satisfied not only by all antieigenvectors but also by all of the eigen-vectors of A.*

In short, the antieigenvalue variational functional $\mu(u) = \mathrm{Re}\,\langle Au, u\rangle/\|Au\|\,\|x\|$ is minimized at antieigenvectors and maximized at eigenvectors.

I have depicted the situation in Fig. 3.2. Given the matrix $A = \begin{bmatrix} 2 & 0 \\ 0 & 1 \end{bmatrix}$, I wanted to portray the full antieigenvalue quotient $\mu(x) = \langle Ax, x\rangle/\|Ax\|\,\|x\|$ for all x. To do so I resorted to letting x's second component $x_2 = \lambda x_1$ where λ is a real parameter to run from $-\infty$ to $+\infty$. Thus we are working in real

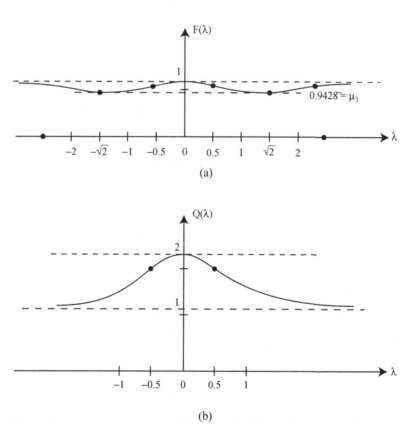

(a)

(b)

Fig. 3.2. The antieigenvalue variational quotient.
(a) $F(\lambda) = (2 + \lambda^2)/(\lambda^4 + 5\lambda^2 + 4)^{1/2}$.
(b) $Q(\lambda) = (2 + \lambda^2)/(1 + \lambda^2)$.

two-space. Then

$$\mu(x) = \frac{\langle Ax, x \rangle}{\|Ax\| \|x\|} = F(\lambda) = \frac{(2 + \lambda^2)}{(\lambda^4 + 5\lambda^2 + 4)^{1/2}}$$

is seen to attain its maximum value 1 at $\lambda = 0$ or $\lambda = \infty$ corresponding to the two eigenvectors, and attains its minimum value $\cos \phi(A)$ at $\lambda = \pm\sqrt{2}$ corresponding to the two antieigenvectors.

Let us elaborate this a bit. For

$$A = \begin{bmatrix} 2 & 0 \\ 0 & 1 \end{bmatrix}, \quad x = \begin{bmatrix} x_1 \\ x_2 \end{bmatrix},$$

let $x = (x_1, x_2)^T = (x_1, \lambda x_1)^T$ with λ a real parameter to allow visualization in two dimensions. Then $\|x\| = (1 + \lambda^2)^{1/2}|x_1|$, $\|Ax\| = (4 + \lambda^2)^{1/2}|x_1|$ and $\langle Ax, x \rangle = (2 + \lambda^2)|x_1|^2$, hence the antieigenvalue variational quotient becomes

$$\mu(\lambda) \equiv F(\lambda) = \frac{(2 + \lambda^2)}{(\lambda^4 + 5\lambda^2 + 4)^{1/2}} = \frac{\text{Re} \langle Ax, x \rangle}{\|Ax\| \|x\|}.$$

We already knew that the first antieigenvalue was

$$\mu_1 = \frac{2\sqrt{2}}{3} \cong 0.9428$$

and to check that variationally on Fig. 3.2, we compute

$$F'(\lambda) = \frac{\lambda^3 - 2\lambda}{(\lambda^4 + 5\lambda^2 + 4)^{3/2}}.$$

From this we see that $F(\lambda)$ takes its minima values μ_1 at $\lambda = \pm\sqrt{2}$, which confirms what we already knew from our antieigenvector expressions

$$x_\pm = \begin{bmatrix} \pm 1 \\ \sqrt{2} \end{bmatrix} \Big/ 3.$$

$F(\lambda)$ takes its maximum values 1 at $\lambda = 0$ and as $\lambda \to \infty$, which correspond to the two eigenvectors $x_2 = [1, 0]^T$ and $x_1 = [0, 1]^T$, respectively.

Just for comparison, I also depicted in Fig. 3.2(b) the easier Rayleigh quotient for the eigenvector theory. Namely, with the same trick

$x = (x_1, \lambda x_1)^T$ used for visualization above, we find

$$Q(\lambda) = \frac{\langle Ax, x \rangle}{\langle x, x \rangle} = \frac{2 + \lambda^2}{1 + \lambda^2} = 1 + \frac{1}{1 + \lambda^2},$$

$$Q'(\lambda) = \frac{-2\lambda}{(1 + \lambda^2)^2},$$

$$Q''(\lambda) = \frac{-2(1 - 3\lambda^2)}{(1 + \lambda^2)^3}.$$

From these one easily draws $Q(\lambda)$ in Fig. 3.2(b). $Q(\lambda)$ achieves its maximum at $\lambda = 0$ and its minimum as $\lambda \to \infty$ corresponding to the two eigenvectors x_2 and x_1. Its inflection points occur at $\lambda = \pm\sqrt{3}/3 \cong 0.57735$.

Exercise 4 at the end of this chapter asks the interested student to confirm some of these calculations and to go further and find the inflection points for the antieigenvalue quotient $F(\lambda)$. Those turn out to occur approximately at $\lambda = \pm0.48382$ and $\lambda = \pm2.29425$.

3.4 Higher Antieigenvalues and Antieigenvectors

Originally (see [Gustafson (1972a)]) I defined "higher" antieigenvalues μ_k and their corresponding higher antieigenvectors $x_\pm^k$ to be those vectors which minimized the antieigenvalue quotient $\mu(u)$ on the subspace orthogonal to all previous antieigenvectors. That was to mimic the usual Rayleigh variational characterizations of "higher" eigenvalues. However, later in [Gustafson (1994b)] I advocated the conceptual advantage of instead taking a combinatorial viewpoint of just minimizing the antieigenvalue quotient $\mu(u)$ over the span of the full eigenvector set minus those eigenvectors which had already been incorporated into the preceding antieigenvectors. Although the two viewpoints often coincide, I like the combinatorial approach better.

For example, suppose you have a 4×4 SPD matrix A. The first antieigenvector pair $x_\pm^1$ will incorporate eigenvectors x_1 and x_4 and the second antieigenvectors pair $x_\pm^2$ will incorporate eigenvectors x_2 and x_3. But as we shall see in some applications later, it is also useful to look at matrix turning angles determined by the subspaces sp$\{x_1, x_3\}$ and sp$\{x_2, x_4\}$. Thus the second, combinatorial thinking, has more flexibility. One may pick out any convenient reducing subspaces of A as one likes, just by selecting eigenvector pairs. You do not need to be orthogonal to anything. Of course for A selfadjoint or normal, you really are orthogonal to the orthonormal complement of this span of the eigenvectors you have chosen.

In particular, higher antieigenvalues and antieigenvectors enter into certain inequalities in matrix statistics in a very natural and essential way. So, to get used to that, let us order our eigenvalues here as is done in statistics, $\lambda_1 \overset{\geq}{=} \lambda_1 \overset{\geq}{=} \cdots \overset{\geq}{=} \lambda_n > 0$ for an SPD matrix A. Then the higher antieigenvalues, higher antieigenvectors and turning angles are defined respectively to be

$$\mu_k(A) = \cos \phi_k(A) = \frac{2\sqrt{\lambda_k \lambda_{n-k+1}}}{\lambda_k + \lambda_{n-k+1}},$$

$$\nu_k(A) = \sin \phi_k(A) = \frac{\lambda_k - \lambda_{n-k+1}}{\lambda_k + \lambda_{n-k+1}},$$

$$x_{\pm}^k = \pm \left(\frac{\lambda_k}{\lambda_k + \lambda_{n-k+1}} \right)^{1/2} x_{n-k+1} + \left(\frac{\lambda_{n-k+1}}{\lambda_k + \lambda_{n-k+1}} \right)^{1/2} x_k.$$

Alternately, if you like to order your eigenvalues the other way around, just be sure to make the numerators positive in the $\sin \phi_k(A)$ expressions! And for $n \times n$ SPD matrices A, you may if you wish not take my shortcut of combinatorial thinking and instead go back to the variational quotient $\mu(x) = \langle Ax, x \rangle / \|Ax\| \|x\|$ and arrive at the above quantities by specifying the increasing orthogonality constraints in analogy to the Rayleigh procedure for eigenvectors.

What happens if A has size $n \times n$ with n odd? By either definition we get the above inward-nesting antieigenvector pairs and then in the last step we must accept an eigenvector. In the original approach of minimizing the variational quotients, this will be the "middle" eigenvector. So you must assign that rule to the combinatorial procedure also. For example if we enlarge Example 1.3 from a 4×4 to the 5×5 matrix

$$\begin{bmatrix} 1 & 0 & 0 & 0 & 0 \\ 0 & 2 & 0 & 0 & 0 \\ 0 & 0 & 5 & 0 & 0 \\ 0 & 0 & 0 & 10 & 0 \\ 0 & 0 & 0 & 0 & 20 \end{bmatrix}$$

then by either approach the antieigenvectors $x_\pm^1$ and $x_\pm^2$ are as before and we have left over $x = [0,0,1,0,0]$, which does not turn at all. Just as with other parts of matrix analysis, symplectic geometry coming to mind, the antieigenvector theory plays out more naturally in even-dimensional spaces.

As I was writing this section with its two often equivalent viewpoints toward the higher antieigenvectors switching between, suddenly a deeper meaning of the distinction occurred to me: Suppose we do not care so much about antieigenvalues and antieigenvectors but are focused instead on the third key entity — the critical operator turning angles $\phi_k(A)$ — then my original variational viewpoint defines those angles by the progressively increasing variational minima $\cos\phi_k(A)$. On the other hand, my combinatorial viewpoint with its two-component nature which emanated from the proof of the Min-Max Theorem defines those angles by the progressively decreasing convex minima $\sin\phi_k(A)$ from the curve $\|\epsilon A - I\|$. I will argue in Sec. 6.4 that these imply a new convex geometry for us to explore in the future.

Taking that point of view here, let us consider the entities

$$\sin\phi_{kj}(A) = \frac{\lambda_k - \lambda_j}{\lambda_k + \lambda_j}$$

where k and j may be arbitrarily chosen, rather than chosen to be nesting inward as in the antieigenvectors. Thus we may consider for all turning vectors

$$x_\pm^{kj} = \pm\left(\frac{\lambda_k}{\lambda_k + \lambda_j}\right)^{1/2} x_j + \left(\frac{\lambda_j}{\lambda_k + \lambda_j}\right)^{1/2} x_k$$

their corresponding turning angles and $\sin\phi_{kj}(A)$, the reducing subspace minima of $\|\epsilon A - I\|$ which will occur at $\epsilon_m^{kj} = 2(\lambda_k + \lambda_j)^{-1}$. Of course we will then have the corresponding μ_{kj} antieigenvalues

$$\mu_{kj} = \cos\phi_{kj}(A) = \frac{2\sqrt{\lambda_j\lambda_k}}{\lambda_k + \lambda_j}.$$

Do these turning vectors also satisfy the Euler equation

$$\frac{A^2 x}{\langle A^2 x, x\rangle} - \frac{2Ax}{\langle Ax, x\rangle} + x = 0$$

with the $x_{\pm}^{kj}$ normalized to norm one, i.e., the contributing eigenvectors $\|x_j\| = \|x_k\| = 1$ so normalized? Of course they do. Just restrict A to the reducing subspace sp$\{x_j, x_k\}$ and then the $x_{\pm}^{jk}$ are the antieigenvectors of that reduced two-dimensional SPD operator A and we know them by the proof of the Min-Max Theorem.

However, it is instructive to verify by direct substitution that all of these combinatorial antieigenvectors $x_{\pm}^{jk}$ directly satisfy the Euler equation. The reader may easily do that in Exercise 5 at chapter's end.

Then even more interesting is to show the converse: that no other such two-component vectors satisfy the Euler equation unless their coefficients are antieigenvector weights. I showed this in [Gustafson (1999e; 2002a; 2006b)] within the context of Lagrange multiplier methods as used in matrix statistics. The application of my antieigenvalue analysis to that field will be treated in Chapter 6.

But here I would like to give a new direct demonstration.

Let $x = c_1 x_1 + c_2 x_2$ where x_1 and x_2 are any norm-one eigenvectors for any two different eigenvalues λ_1 and λ_2, and where $|c_1|^2 + |c_2|^2 = 1$. Then we have

$$Ax = c_1 \lambda_1 x_1 + c_2 \lambda_2 x_2,$$

$$A^2 x = c_1 \lambda_1^2 x_1 + c_2 \lambda_2^2 x_2,$$

$$\langle Ax, x \rangle = |c_1|^2 \lambda_1 + |c_2|^2 \lambda_2,$$

$$\langle A^2 x, x \rangle = |c_1|^2 \lambda_1^2 + |c_2|^2 \lambda_2^2.$$

Putting these into the Euler equation produces

$$c_1 \left\{ \frac{\lambda_1^2}{[|c_1|^2 \lambda_1^2 + |c_2|^2 \lambda_2^2]} - 2 \frac{\lambda_1}{[|c_1|^2 \lambda_1 + |c_2|^2 \lambda_2]} + 1 \right\} x_1$$

$$+ c_2 \left\{ \frac{\lambda_2^2}{[|c_1|^2 \lambda_1^2 + |c_2|^2 \lambda_2^2]} - 2 \frac{\lambda_2}{[|c_1|^2 \lambda_1 + |c_2|^2 \lambda_2]} + 1 \right\} x_2 = 0.$$

Because x_1 and x_2 are linearly independent, each of their coefficients in the above must be zero. Suppose $c_1 = 0$. Then the second coefficient becomes

$$\frac{1}{|c_2|^2} - \frac{2}{|c_2|^2} + 1 = 0$$

and hence $|c_2|^2 = 1$. That is just the Euler equation being satisfied by the eigenvector x_2 plus allowing for a phase change of modulus one. Similarly for the $c_2 = 0$ case. We already knew all eigenvectors satisfied the Euler equation.

So let us turn to the more interesting case when both c_1 and c_2 do not vanish. One may proceed in several ways but I found the following calculation the most revealing, in that certain related spectral entities enter in interesting ways. As a hint, remember that the condition number κ of a SPD matrix with eigenvalues $0 < \lambda_1 \overset{\leq}{=} \cdots \overset{\leq}{=} \lambda_n$ is defined to be

$$\kappa = \frac{\lambda_n}{\lambda_1}.$$

In our case because we are letting λ_2 be any other eigenvalue, $\kappa = \lambda_2/\lambda_1$.

Let us now consider when the x_1 coefficient above will vanish: the x_2 coefficient will yield to the same analysis in the same way. Clearing the denominator leads to

$$\lambda_1^2[|c_1|^2\lambda_1 + |c_2|^2\lambda_2] - 2\lambda_1[|c_1|^2\lambda_1^2 + |c_2|^2\lambda_2^2]$$
$$+ [|c_1|^2\lambda_1^2 + |c_2|^2\lambda_2^2] \cdot [|c_1|^2\lambda_1 + |c_2|^2\lambda_2] = 0.$$

Collecting $|c_1|^2$ and $|c_2|^2$ terms gives

$$|c_1|^2\{\lambda_1^3 - 2\lambda_1^3\} + |c_2|^2\{\lambda_1^2\lambda_2 - 2\lambda_1\lambda_2^2\} + |c_1|^4\{\lambda_1^3\}$$
$$+ |c_1|^2|c_2|^2\{\lambda_1^2\lambda_2 + \lambda_1\lambda_2^2\} + |c_2|^4\{\lambda_2^3\} = 0.$$

Putting in the constraint $|c_2|^2 = 1 - |c_1|^2$ and for simplicity of notation calling $|c_1|^2 = x$ leads to the equation

$$x(-\lambda_1^3) + (1 - x)\{\lambda_1^2\lambda_2 - 2\lambda_1\lambda_2^2\} + x^2\{\lambda_1^3\}$$
$$+ x(1 - x)\{\lambda_1^2\lambda_2 + \lambda_1\lambda_2^2\} + (1 - x)^2\{\lambda_2^3\} = 0.$$

Collecting x power terms gives after algebraic simplification

$$x^2\{\lambda_1^3 + \lambda_2^3 - \lambda_1^2\lambda_2 - \lambda_1\lambda_2^2\} + x\{-\lambda_1^3 - 2\lambda_2^3 + 3\lambda_1\lambda_2^2\}$$
$$+ \{\lambda_1^2\lambda_2 - 2\lambda_1\lambda_2^2 + \lambda_2^3\} = 0.$$

Now divide by λ_1^3 and recall $\lambda_2/\lambda_1 = \kappa$, the relative condition number. This brings us to the quadratic equation

$$x^2\{\kappa^3 - \kappa^2 - \kappa + 1\} + x\{-2\kappa^3 + 3\kappa^2 - 1\} + \{\kappa^3 - 2\kappa^2 + \kappa\} = 0.$$

At this point it occurred to me to try to factor the a, b, c coefficients of this quadratic $ax^2 + bx + c = 0$, and lo and behold,

$$a = (\kappa - 1)^2(\kappa + 1),$$

$$b = (-1)(\kappa - 1)^2(2\kappa + 1),$$

$$c = \kappa(\kappa - 1)^2.$$

Since $\kappa > 1$ by $\lambda_2 > \lambda_1$, the quadratic becomes the simplified

$$(\kappa + 1)x^2 + (-1)(2\kappa + 1)x + \kappa = 0.$$

Therefore

$$x = \frac{(2\kappa + 1) \pm \sqrt{(4\kappa^2 + 4\kappa + 1) - 4\kappa(\kappa + 1)}}{2(\kappa + 1)}$$

$$= \frac{(2\kappa + 1) \pm 1}{2(\kappa + 1)}.$$

The "+" root is $x = 1 = |c_1|^2$, the eigenvector case we already disposed of above. The more interesting "$-$" root is

$$x = \frac{2\kappa}{2(\kappa + 1)} = \frac{\kappa}{\kappa + 1} = \frac{\lambda_2 \lambda_1}{1 + \lambda_2/\lambda_1} = \frac{\lambda_2}{\lambda_2 + \lambda_1}$$

which is the $|c_1|^2$ coefficient for an antieigenvector

$$x = \left(\frac{\lambda_2}{\lambda_1 + \lambda_2}\right)^{1/2} x_1 \pm \left(\frac{\lambda_1}{\lambda_1 + \lambda_2}\right)^{1/2} x_2.$$

Thus it has been shown: The only one- or two-component solutions to the Euler equation are the eigenvectors and all of the combinatorial antieigenvectors.

Note that these include all combinations, not just those nesting inward on the spectrum. And that infers interesting skew antieigenvalues $\mu_{jk} = \cos_{jk}(A)$ in the other variational way to approach antieigenvectors.

Each antieigenvector pair $x_\pm$ has its own interesting internal geometrical features. The following was shown in [Gustafson (2000d)]:

Antieigenvector pair angle. Given any antieigenvector pair $x_\pm^{jk}$, then the cosine of the angle between them is

$$\langle x_+^{jk}, x_-^{jk} \rangle = -\sin \phi_{jk}(A).$$

Thus the antieigenvector pair angle is always $\phi_{jk}(A) + \pi/2$.

I have taken the two-component antieigenvector viewpoint throughout this section. Indeed, that is how I think, for almost all of the interesting real applications to-date for other areas of science, engineering, or physics have concerned $n \times n$ positive definite symmetric or Hermitian matrices A. But when one goes beyond selfadjoint to normal matrices, the cup minimum may occur for $\|\epsilon A - I\|$. Here is an example. Let

$$A = \begin{bmatrix} 1+i & 0 & 0 \\ 0 & 2+i & 0 \\ 0 & 0 & 1+2i \end{bmatrix} = \begin{bmatrix} \lambda_1 & 0 & 0 \\ 0 & \lambda_2 & 0 \\ 0 & 0 & \lambda_3 \end{bmatrix}.$$

A is normal and strongly accretive. Its numerical range $W(A)$ is the convex hull triangle with corners its three eigenvalues in the complex plane. Because $\epsilon A - I$ is also normal, its norm $\|\epsilon A - I\|$ is the same as its spectral radius $|\sigma(\epsilon A - I)|$. Thus

$$\|\epsilon A - I\| = \max\{|\epsilon \lambda_1 - 1|, |\epsilon \lambda_2 - 1|, |\epsilon \lambda_3 - 1|\}.$$

A simple calculation reveals that λ_3 always wins for all $\epsilon > 0$. Therefore easy calculations from $|\epsilon \lambda_3 - 1|^2 = 5\epsilon^2 - 2\epsilon + 1$ show that

$$\sin^2 \phi(A) = \|\epsilon_m A - I\|^2 = \frac{4}{5}$$

occurs at $\epsilon_m = 1/5$ and that will be a cup minimum.

The point is that for nonselfadjoint A, the entity $\|\epsilon A - I\|$ should be extended to $\|zA - I\|$ with z ranging over the whole complex plane. Although we have done that to some extent for normal matrices, e.g., see [Gustafson (2005b, 2010b)], generally I have not been too interested because I have not seen any physical application to drive such an investigation.

Finally, one should ask whether three or higher component vectors x might enter interestingly. I leave this issue partially open. Exercise 6 suggests the student might have fun putting a three-component x into my calculation above and observe what happens. In [Gustafson (1993a)] authored with M. Seddighin, we showed rather generally for normal operators that, from the variational quotient viewpoint, all antieigenvectors are of two- or one-component. For this see also the book of [Gustafson (1997a)] with D. Rao. But I do not believe all has been discovered yet. If you think of my antieigenvalue variational quotient as a surface within a new kind of Morse theory, one could have valleys, saddle points, and the like. In other words, there could be further interesting new operator trigonometry beyond just looking at local minima and maxima on reducing subspaces.

3.5   The Triangle Inequality

Early on D. Rao and I proved the following.

Theorem 3.3 (The Triangle Inequality). *Let x, y, z be three unit vectors in a Hilbert space. Let angles $\phi_{xy}, \phi_{yz}, \phi_{xz}$ be defined by $\cos \phi_{xy} = \operatorname{Re} \langle x, y \rangle$, $\cos \phi_{yz} = \operatorname{Re} \langle y, z \rangle$, $\cos \phi_{xz} = \operatorname{Re} \langle x, z \rangle$, with $0 \overset{\leq}{=} \phi_{xy}, \phi_{yz}, \phi_{xz} \overset{\leq}{=} \pi$. Then*

$$\phi_{xz} \overset{\leq}{=} \phi_{xy} + \phi_{yz}.$$

This inequality was stated without proof in [Krein (1969)] and indeed is intuitively obvious. However, we found no trivial proof and showed it essentially as follows, see [Rao (1972)] or our joint paper [Gustafson (1977a)]. It is sufficient to prove that

$$\cos \phi_{xz} \overset{\geq}{=} \cos(\phi_{xy} + \phi_{yz}) = \cos \phi_{xy} \cos \phi_{yz} - \sin \phi_{xy} \sin \phi_{yz}.$$

Moving the term $\sin \phi_{xy} \sin \phi_{yz}$ to the left and squaring brings us to the sufficient condition

$$\sin^2 \phi_{xy} \sin^2 \phi_{yz} \overset{\geq}{=} (\cos \phi_{xy} \cos \phi_{yz} - \cos \phi_{xz})^2$$

which becomes an important inequality that will appear elsewhere in this book

$$1 - \cos^2 \phi_{xy} - \cos^2 \phi_{yz} - \cos^2 \phi_{xz} + 2 \cos \phi_{xy} \cos \phi_{yz} \cos \phi_{xz} \overset{\geq}{=} 0. \quad (***)$$

But that inequality follows from the positive semi-definiteness of the Gram matrix

$$G(x, y, z) = \begin{bmatrix} \langle x, x \rangle & \langle x, y \rangle & \langle x, z \rangle \\ \langle y, x \rangle & \langle y, y \rangle & \langle y, z \rangle \\ \langle z, x \rangle & \langle z, y \rangle & \langle z, z \rangle \end{bmatrix}$$

and its complex conjugate $\overline{G(x, y, z)}$ from which we get the positive semi-definite real matrix

$$\begin{bmatrix} 1 & \cos \phi_{xy} & \cos \phi_{xz} \\ \cos \phi_{xy} & 1 & \cos \phi_{yz} \\ \cos \phi_{xz} & \cos \phi_{yz} & 1 \end{bmatrix}$$

whose nonnegative determinant is the above inequality.

Krein (1969) also stated the following triangle inequality for invertible operators on a Hilbert space

$$\phi(AB) \overset{\leq}{=} \phi(A) + \phi(B),$$

here stated in my notation. This follows directly from the above more fundamental vector triangle inequality. One has for every x

$$\phi_{x,ABx} \overset{\leq}{=} \phi_{x,A^{-1}x} + \phi_{A^{-1}x,ABx}$$

$$= \phi_{x,Ax} + \phi_{x,Bx}$$

so their suprema give you the operator triangle inequality above.

The latter triangle inequality for operators or matrices is expected and nice to have. My own interest returned to the more fundamental vector turning-angle triangle inequality (∗∗∗) above when I applied my operator trigonometry to quantum mechanics. That will be discussed in Chapter 7. There I came to the opinion that even more fundamental is the more general inequality

$$1 + 2a_1 a_2 a_3 - (a_1^2 + a_2^2 + a_3^2) \overset{\geq}{=} 0$$

obtained from the Gram matrix determinant

$$|G| = \begin{vmatrix} 1 & a_1 & a_3 \\ a_1 & 1 & a_2 \\ a_3 & a_2 & 1 \end{vmatrix} \overset{\geq}{=} 0$$

which I have written in a more general form than that for cosines above. This will be discussed further in Chapter 7.

Then in the intervening years I have found a very interesting connection of inequality (∗∗∗) to geometric invariant theory. Moreover just in the last year a similar inequality used by matrix statisticians has come to my attention. These important connections will be spelled out in a paper under preparation [Gustafson (2011c)].

3.6 Extended Operator Trigonometry

In [Gustafson (2000d)] I developed one way to extend my operator trigonometry to arbitrary invertible matrices and operators. One could further extend to noninvertible operators by the usual quotienting out of the null space $N(A)$, but usually for simplicity I just assume invertible operators.

For invertible matrices A one immediately thinks of using the singular value decomposition

$$A = U\Sigma V^*$$

and indeed one could proceed in that way but I have not done so: I would need a guiding application of some importance to motivate me. Instead I found myself going to the left polar form

$$A = U|A|$$

where $|A|$ is the absolute value operator $(A^*A)^{1/2}$. For invertible A I can then revert the general operator trigonometry to that already known for the positive selfadjoint operator $|A|$ by adapting my $\sin\phi(A)$ norm curve to the norm curve

$$\sin\phi_e(A) = \inf_\epsilon \|\epsilon A - U\|.$$

From this one has for matrices A that

$$\sin\phi_e(A) = \min_{\epsilon>0} \|\epsilon|A| - I\|$$
$$= \frac{\sigma_1(A) - \sigma_n(A)}{\sigma_1(A) + \sigma_n(A)}$$

where the σ's are A's singular values. Moreover I may recover the important Min-Max Theorem by adapting $\cos\phi(A)$ to

$$\cos\phi_e(A) = \inf_{x\neq 0} \frac{\langle|A|x,x\rangle}{\||A|x\|\,\|x\|} = \inf_{x\neq 0} \frac{\langle Ax, Ux\rangle}{\|Ax\|\,\|x\|}$$

from which for matrices A we have

$$\cos\phi_e(A) = \frac{2\sqrt{\sigma_1(A)\sigma_n(A)}}{\sigma_1(A) + \sigma_n(A)}.$$

There are two features I particularly like about using polar form rather than singular value decomposition. First, polar form holds more generally, not only for matrices but for all bounded operators A on a Hilbert space, even for closed densely defined A there. Secondly the unitary U in $A = U|A|$ removes the uniform rotational factor of A, which has nothing to do with my concepts of operator turning angles. I repeat this last statement because it seems that many do not understand my operator trigonometry because upon encountering it for the first time their minds automatically go for example to "group representations" with its unitary rotations, or the like.

In the endnotes to Chapter 3 in the book [Gustafson (1997a)] I wax philosophic about the names *eigenvector* and *antieigenvector* and suggest a better name for the latter might be *winkelvectors*, i.e., anglevectors. In view of the paragraph above, I would now prefer the name *twistvectors*. That is a better way to physically visualize my operator trigonometry. But the term *antieigenvectors* still has its merit of contrast to the term *eigenvector* which means no turning (or twisting) at all.

Commentary

Although antieigenvalues and antieigenvectors are my creation, a notion of operator angle was defined independently by M. G. Krein (1969), and also in a partial sense by H. Wielandt (1967). My Ph.D. student D. Rao discovered in 1971 the Krein paper, which appeared slightly after my earlier 1967 and 1968 notices and papers. However it came from quite different motivations. Krein was interested in bounding the spectrum of an exponential integral in the complex plane. He called his operator angle dev(A) and it is equivalent to my operator angle $\phi(A)$. Wielandt's Wisconsin 1967 Lecture notes were first pointed out to me by H. Schneider in 1994. In them he defined what he called the singular angle $\gamma(A)$ for square matrices A. His motivation was to use it to study the spectrum $\sigma(A + B)$ in terms of the spectra of $\sigma(A)$ and $\sigma(B)$. I have given a full account of the relationships of their ideas to mine in [Gustafson (1996c)].

The bottom line is that because they were not led as I was to need and define $\sin \phi(A)$ in terms of the convex curve $\|\epsilon A - I\|$ minimum, they were not led to the other half of what became my full operator trigonometry. Also their contexts did not lead them to my variational Rayleigh-type quotient $\mu(A)$, which I minimized at antieigenvectors and maximized at eigenvectors. Thus they also had no Euler equation.

My 1968 Min-Max Theorem was later proved in a different way by Asplund and Ptak (1971). I was, speaking frankly here, disappointed that Asplund, who had actively and enthusiastically discussed with me my Min-Max Theorem at the 1969 Los Angeles *Third Symposium on Inequalities* after I presented it there, did not cite my priority. I told him in particular that I was going to explore whether it extended to Banach space, and indeed

I say that in my 1969 Symposium paper which appeared later [Gustafson (1972a)].

Asplund and Ptak (1971) showed that my Min-Max result fails, for some operator pair A, B, in every Banach space. That is a valuable result. However, I must also add that when you look closely at their paper, you will not find the proof to be as constructive as mine. Thus their approach does not naturally give anything conceptual or practical about antieigenvalues or antieigenvectors. But it saved me from looking into Banach space as concerns my Min-Max Theorem.

Chandler Davis invited me to Toronto in 1981 and showed me his work [Davis (1980)] extending the Kantorovich (1948) inequality to normal matrices. I had not previously been aware of the Kantorovich inequality. Davis deserves credit for extending the particulars of my antieigenvalue theory for SPD matrices to normal matrices. Also he showed me the paper of Mirman (1983) which uses nice convexity and numerical range techniques to calculate antieigenvalues.

3.7 Exercises

1. Prove the Toeplitz–Hausdorff Theorem: The numerical range $W(A)$ of any linear operator in any real or complex inner product is convex.
2. Work through the paper [Gustafson (1968c)] until you find a mathematical typo. This typo has no mathematical consequences to the rest of the paper, however. Read further about the geometry of Banach space, using the citations of that paper and the additional ones given in Sec. 3.1.
3. The norm curve $\|\epsilon B - I\|$ has a unique minimum in uniformly convex Banach spaces. But this is not true in general Banach spaces. Construct a counterexample.
4. Verify the calculations for $F'(\lambda)$ for the example in Fig. 3.2, where $F(\lambda)$ is the antieigenvalue functional. Then calculate $F''(\lambda)$ and find the inflection points of the $F(\lambda)$ curve as depicted in Fig. 3.2. Confirm that $F(\lambda) \to 1$ as $\lambda \to \pm\infty$.
5. Show by direct substitution that antieigenvectors

$$x_{\pm}^{jk} = \pm \left(\frac{\lambda_k}{\lambda_k + \lambda_j}\right)^{1/2} x_j + \left(\frac{\lambda_j}{\lambda_k + \lambda_j}\right)^{1/2} x_k$$

satisfy the Euler equation

$$\frac{A^2x}{\langle A^2x, x\rangle} - 2\frac{Ax}{\langle Ax, x\rangle} + x = 0.$$

6. Play with the Euler equation with three-component vectors $x = c_1x_1 + c_2x_2 + c_2x_3$ where the x_i are eigenvectors. Go further to such trial vectors with k components.

Chapter 4

Applications in Numerical Analysis

Perspective

The scientific needs of the country along with my conscience returned my interest to applied mathematics in 1980. That year I resurrected the Ph.D. in Applied Mathematics within our Department of Mathematics. I also entered into a number of very stimulating NSF-funded projects with engineering teams in aeronautical, chemical, civil, and electrical and computer engineering. Thus my principal mathematical preoccupations in the decade of the 1980s centered upon those engineering tasks. You will find very, very few citations to works by me on my antieigenvalue theory in the 1970s and 1980s. If you are doing physics or engineering in a real way, for example if you want to really contribute to fluid dynamics, you better be thinking real fluid dynamics. You must become as competent and knowledgeable as your coworkers from those applied fields. That is my opinion.

See for example our book [Gustafson (1990a)] and conference volume [Gustafson (1991b)].

As a result, all that computational experience gained in the 1980s accrued in an essential way to enable my applying of my antieigenvalue theory to numerical methods when I returned to my operator trigonometry in the 1990s. As I mentioned in the Preface to this book, the first opportunity to do so came with the invitation to speak at the Dubrovnik Workshop in June 1990. In computational work on the equations of elasticity with my Ph.D. student N. Sobh in the late 1980s, we had employed preconditioned conjugate gradient methods (see our paper [Gustafson (1991c)]). That brought my attention to the Kantorovich convergence rate bounds for such methods. That such bounds are instantly within my antieigenvalue theory will be recounted in Sec. 4.1.

Also in the 1980s I had been doing a lot of multigrid and domain decomposition computation for aerodynamics with my Ph.D. student R. Leben (see for example our paper [Gustafson (1988a)]). Even earlier with my Ph.D. students D. P. Young, H. Tadjeran, B. Eaton, R. Hartman, K. Halasi, and E. Ash I had become familiar with ODE solvers, PDE minimum residual solvers, relaxation schemes such as SOR, and fast Poisson Solvers in general. Later with my Ph.D. student J. McArthur we would add the ADI schemes to our repertoire. See for example our papers [Gustafson (1981b; 1982a; 1983b; 1985a; 1990b; 1992a)].

From that acquired acumen in linear solver theory and practice I was able in the 1990s to bring such schemes within the purview of my operator trigonometry. My results for minimum residual schemes, relaxation schemes, ADI, domain decomposition and multilevel schemes, the Model Problem, and preconditioning methods will be recounted in Secs. 4.2–4.6. I must assume the reader without much background in computational linear algebra will be willing to read further from the bibliographies within the cited papers.

4.1 Gradient Descent: Kantorovich Bound is Trigonometric

In the fluid dynamics linear solver literature and also in the economics literature you will find the fundamental Kantorovich bound for steepest descent

$$E(x_{k+1}) \overset{\leq}{=} \left(1 - \frac{4\lambda_1 \lambda_n}{(\lambda_1 + \lambda_n)^2}\right) E(x_k).$$

Here $E(x)$ is the error $\langle (x - x^*), A(x - x^*)\rangle /2$, x^* the true solution. See for example the books [Golub and Van Loan (1987); Luenberger (1973)].

I knew instantly from my antieigenvalue theory that the content of this bound was exactly operator-trigonometric:

$$E(x_{k+1}) \overset{\leq}{=} \sin^2 \phi(A) E(x_k).$$

Nowhere in the numerical nor in the optimization literature did I find this important geometrical result: that gradient and conjugate gradient convergence rates are fundamentally trigonometric. That is, they are a direct reflection of the elemental fact of A's maximum turning angle $\phi(A)$.

The known conjugate gradient convergence rate, trigonometrically written, becomes,

$$E(x_k) \overset{<}{=} 4(\sin \phi(A^{1/2}))^{2k} E(x_0).$$

See [Gustafson (1994a)] where I published this result for conjugate gradient convergence. And I first announced the steepest descent result at the 1990 Dubrovnik conference [Gustafson (1991a)].

As I related in the commentary at the end of the previous chapter, C. Davis had shown me another version of the Kantorovich inequality in 1980 in Toronto. But I did not pay much attention. It was only when we were actually computing Navier elasticity in the late 1980s as reported in [Gustafson (1991c)] that the full importance of the above bound for iterative linear solver theory hit me. It is so beautiful . . . it is immediately

$$E(x_{k+1}) \overset{<}{=} (1 - \cos^2 \phi(A)) E(x_k).$$

For the first time I realized that I had to return to my original antieigenvalue operator trigonometry days of twenty years earlier and bring them into the light of day by showing their application to numerical analysis and later to statistics, quantum mechanics, and recently, financial instruments. Thus as I stated in the Preface to this book, 1990 and the somewhat improbable invitation to speak at the Dubrovnik conference signals my return to more fully develop my antieigenvalue analysis during the last twenty years.

Later when H. Schneider showed me Wielandt's Wisconsin lecture notes and asked me to write the paper [Gustafson (1996c)] in which I compared my ideas with those of Wielandt and Krein, I was led to also compare the so-called Kantorovich–Wielandt inequality of subspace theory to my operator trigonometry. I state the result in [Gustafson (1996c)] as a footnote on p. 362 there. The Kantorovich–Wielandt inequality, e.g., see the book [Horn and Johnson (1985)], begins by defining an angle θ in the first quadrant in terms of the usual matrix condition number κ, according to

$$\cot\left(\frac{\theta}{2}\right) = \kappa.$$

Recall that $\kappa = \lambda_n/\lambda_1$ or more generally σ_1/σ_n in terms of the matrix A's eigenvalues or singular values, respectively. Then the K–W inequality states that

$$|\langle Ax, Ay \rangle| \overset{<}{=} \cos\theta \|Ax\| \|Ay\|$$

for every pair of orthogonal vectors x and y. My result of [Gustafson (1996c)], later expanded in [Gustafson (1999b)], shows that the relationship of the K–W angle θ to my antieigenvalue theory is precisely

$$\cos \phi(A^*A) = \sin \theta(A).$$

Stated another way, the sine of the K–W angle is precisely the first antieigenvalue of A^*A.

Much of numerical linear algebra has been strongly conditioned to think always in terms of the condition number κ, and the K–W inequality is no exception. With my operator trigonometry we have now a much better window into the actual dynamics of a matrix transformation $x \to Ax$. As I point out in [Gustafson (1999b)], anytime for an SPD matrix A you encounter some entity γ in your theory for which

$$\gamma = \frac{\kappa - 1}{\kappa + 1}$$

then we have a three-way connection between your theory, the Kantorovich–Wielandt theory, and my direct operator trigonometry, given by

$$\gamma = \cos \theta(A^{1/2}) = \sin \phi(A).$$

4.2   Minimum Residual $Ax = b$ Solvers

At the Iterative Methods conference at Breckenridge, Colorado, in 1994 I went further [Gustafson (1994c)] and announced new trigonometric understandings of several iterative linear solvers for $Ax = b$. Those included GCR(k), Orthomin, CGN, GMRES, CGS, among others. However the referees of my conference paper did not seem to understand my geometrical concepts. Therefore I more adequately exposited those concepts in more detail in a series of papers from 1996 to 2004. Those results will be presented in the rest of this chapter. I will give here only representative results and citations to the papers. It would require a full book to treat completely all the needed background and contents of computational linear algebra within which my operator trigonometric interpretations take place.

Beyond the conference paper [Gustafson (1996b)] I would suggest beginning with the journal paper [Gustafson (1997b)]. Here is a representative trigonometric result. The GCR residual error bound

$$\|r_i\| \overset{\leq}{=} \min_{q_i \in P_i} \|q_i(A)\| \|r_0\| \overset{\leq}{=} \left[1 - \frac{(\lambda_{\min}(\text{Re}\,(A))^2}{\lambda_{\max}(A^T A)} \right]^{1/2} \|r_0\|$$

in terms of polynomials $q_i(A)$, can be trigonometrically improved to the convergence rate $[1 - \cos^2 \phi(A)]^{1/2} = \sin \phi(A)$. Always in such improvements one must assure that one's hypotheses place you in the intersection of results from the numerical theory with the assumptions of my operator-trigonometric theory. A usually safe rule-of-thumb is to take A to be SPD.

I mention a few more improved error estimates to minimum residual schemes via my operator trigonometry in the later paper [Gustafson (2003a)]. I am currently going back to the important GMRES algorithm for a closer look, among other things I am doing at the present time.

4.3   Richardson Relaxation Schemes (e.g., SOR)

In [Gustafson (1997b)] I also look operator-trigonometrically at a number of other iterative solvers. Let the general Richardson iteration be defined as

$$x_{k+1} = x_k + \alpha(b - Ax_k)$$

with iteration matrix $G_\alpha = I - \alpha A$ and convergence factor $\rho(I - \alpha A)$. Let A be strongly accretive.

Theorem 4.1 (Richardson is Trigonometric). *The optimal parameter α_{opt} and consequent optimal convergence factor ρ_{opt} for Richardson iteration are, respectively,*

$$\alpha_{\text{opt}} = \epsilon_m(A) = \frac{\text{Re}\,\langle Ax_+, x_+ \rangle}{\|Ax_+\|^2}$$

and

$$\rho_{\text{opt}} = \|\epsilon_m A - I\| = \sin \phi(A),$$

where x_+ is a first antieigenvector of A.

See [Gustafson (1997b)] for the proof and details. Also there you will find new operator trigonometry of the Uzawa, Chebyshev, Jacobi, Gauss–Seidel, SOR, SSOR schemes. I also treat the meaning of superlinear convergence of conjugate gradient methods, and give a preliminary trigonometric treatment of the AMLI multilevel methods.

In the subsequent paper [Gustafson (1998g)] I extended our trigonometric understandings of such schemes by working out many details for the so-called Model Problem

$$
\begin{cases}
-\Delta u = f(x, y) & \text{in } \Omega, \\
 u = g(x, y) & \text{in } \partial\Omega,
\end{cases}
$$

where Ω is the square $0 < x < 1, 0 < y < 1$ and Δ is the Laplacian operator discretized by centered differences on a uniform mesh of size $h = 1/N$. Here are some of my findings.

Let A_h denote the discretized Laplacian in the usual matrix form. Then the operator angle $\phi(A_h)$ is exactly

$$
\phi(A_h) = \left(\frac{\pi}{2}\right) - \pi h.
$$

Moreover the two principal operator trigonometric entities for this discrete Laplacian are

$$
\sin \phi(A_h) = \cos \pi h,
$$

$$
\cos \phi(A_h) = \sin \pi h.
$$

The first antieigenvectors are

$$
x_\pm = \pm \cos\left(\frac{\pi h}{2}\right) e_{11} + \sin\left(\frac{\pi h}{2}\right) e_{N-1}e_{N-1}
$$

where of course e_{11} and $e_{N-1}e_{N-1}$ denote the lowest and highest eigenvectors of the discretized problem.

As to particular schemes: I find the spectral radius of the Jacobi iteration matrix B to be $\sin \phi(A_h)$. The Gauss–Seidel iteration matrix $\mathcal{L}$ has spectral radius $\sin^2 \phi(A_h)$.

This paper [Gustafson (1998g)] led me to a deeper interest in the venerable SOR method of D. M. Young. In particular I show that the optimal spectral radius for the pointwise SOR iteration matrix $\mathcal{L}_\omega$ is

$$
\rho(\mathcal{L}_{\omega_{\text{opt}}}^{\text{SOR}}) = \frac{1 - \cos \phi(A_h)}{1 + \cos \phi(A_h)}.
$$

The optimal relaxation parameter, so important in real applications of SOR, is

$$\omega_{\text{opt}} = \frac{2}{1 + \cos\phi(A_h)}.$$

I returned to SOR in the later paper [Gustafson (2003a)]. I think that we have barely scratched the surface of the rich trigonometric content of SOR. In particular, I show here in Fig. 4.1 the famous D. M. Young picture (1971) of the SOR spectral radius $s(\mathcal{L}_\omega)$ versus relaxation parameter ω, juxtaposed with my operator trigonometric convex operator norm curve $\|\epsilon A^{1/2} - I\|$ versus my parameter ϵ.

Recall that in the SOR scheme one writes A in its diagonal and upper and lower triangular parts

$$A = D - E - F$$

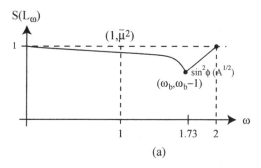

(a)

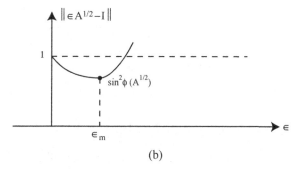

(b)

Fig. 4.1. SOR and operator trigonometry.
(a) The famous D. M. Young picture of SOR.
(b) The equivalent picture, seen operator trigonometrically.

and then arrives at the iteration matrix

$$\mathcal{L}_\omega = (1 - \omega D^{-1}E)^{-1}[(1 - \omega)I + \omega D^{-1}F].$$

For a large class of operators the iterative convergence is optimized at relaxation parameter

$$\omega_b = \frac{2}{1 + (1 + \bar{\mu}^2)^{1/2}}$$

where $\bar{\mu} = S(B)$ is the spectral radius of the corresponding Jacobi iteration matrix for A. In [Young (1971), p. 201, Fig. 5.1], the Dirichlet Model Problem I described above with $h = 1/20$ is treated by SOR and the resulting behavior of the iteration matrix spectral radius $S(\mathcal{L}_\omega)$ as a function of relaxation parameter ω is sketched as I did here in Fig. 4.1(a). The optimal ω occurs at $\omega = 1.72945$ and that is denoted by ω_b in Fig. 4.1(a). As can be vividly seen from the figure, Young's curve is optimized (minimized) at the relaxation value ω_{opt} which I stated above in trigonometric terms, and achieves the height $\sin^2 \phi(A^{1/2})$ there. Note from Fig. 4.1(b) that my operator curve achieves the same optimal (minimal) value at $\epsilon_m = 2/(\lambda_1 + \lambda_n)$. The general relationship between Young's relaxation parameter and my ϵ parameter now also regarded as a relaxation parameter is

$$\epsilon_m(A)\omega_{\text{opt}}(A) = (\epsilon_m(A^{1/2}))^2.$$

This research brought out the following general interesting operator-trigonometric identity.

Lemma 4.1 (Square Root Identity). *Let A be an SPD matrix. Then*

$$\sin \phi(A^{1/2}) = \frac{\sin \phi(A)}{1 + \cos \phi(A)} = \tan\left(\frac{\phi(A)}{2}\right).$$

Generally speaking, the operator trigonometry will have its own set of identities.

4.4 Very Rich Trigonometry Underlies ADI

The paper [Gustafson (1998g)] uncovered an incredibly rich operator trigonometry of the highly successful ADI (Alternating-Direction-Implicit, also called the Peaceman–Rachford scheme) algorithm which to my impressions is still employed in refined forms in both commercial and NSA CFD software packages. Following the treatment of ADI in [Varga (1962); Axelsson (1994)] I found the following amazing trigonometric convergence rate bounds $d(\alpha, \beta, q)$ for ADI when employed on the Model Problem with semi-iteratively improved optimizing convergence parameters

$$d(\alpha, \beta, 1) = \frac{1 - \sqrt{\alpha/\beta}}{1 + \sqrt{\alpha/\beta}} = \sin \phi(A_h)^{1/2},$$

$$d(\alpha, \beta, 2) = \frac{1 - (\cos \phi(A_h))^{1/2}}{1 + (\cos \phi(A_h))^{1/2}},$$

$$d(\alpha, \beta, 4) = \cdots = \frac{1 - (\cos \phi(A_h))^{1/4}/\cos \left(\phi(A_h)/2\right)}{1 + (\cos \phi(A_h))^{1/4}/\cos \left(\phi(A_h)/2\right)}.$$

For the coarse mesh example of [Gustafson (1998g)]

$$A = \frac{1}{9} \begin{bmatrix} 4 & -1 & -1 & 0 \\ -1 & 4 & 0 & -1 \\ -1 & 0 & 4 & -1 \\ 0 & -1 & -1 & 4 \end{bmatrix}$$

where $N = 3$ and $h = 1/3$ for the Model Problem, these semi-iterative convergence rate numbers are easily calculated from $30°$ and $15°$ to be

$$d_1 = 0.268,$$
$$d_2 = 0.036,$$
$$d_3 = 0.00065.$$

Such dramatic semi-iterative improvement in ADI is indicative of why it is still such an important method for simulations in two- and three-space dimensions. However, it is forbidding to try to delve deeper into its possible further underpinning operator-trigonometric features, especially in ADI's noncommutative versions, and I have not done so.

4.5 Domain Decomposition Multilevel Schemes

As the numerical methods become more intricate and as they combine a number of hybrid features, it becomes more difficult to see their possible hidden operator-trigonometric dynamics. Nonetheless, there are some.

In my first papers [Gustafson (1996b; 1997b)] I considered the AMLI (Algebraic Multilevel Iteration) scheme for a nested sequence of finite element meshes

$$\Omega_\ell \supset \Omega_{\ell-1} \supset \cdots \supset \Omega_{k_0}$$

with finite element vector spaces upon them for a second-order elliptic boundary value problem. Suppose V_1 and V_2 be disjoint finite element subspaces of a given level V_ℓ and let $a(u,v)$ be the $H^1(\Omega)$ SPD energy inner product. Then

$$\gamma = \sup_{u \in V_1, v \in V_2} \frac{a(u,v)}{(a(u,u) \cdot a(v,v))^{1/2}}$$

is the cosine of the angle between the subspaces V_1 and V_2 in the $a(u,v)$ inner product. One is led to a Schur 2×2 matrix decomposition

$$A = \begin{bmatrix} A_{11} & A_{12} \\ A_{21} & A_{22} \end{bmatrix}$$

with spectral radius

$$\rho\left(A_{11}^{-1/2} A_{12} A_{22}^{-1/2}\right) = \gamma.$$

Let D be the block-diagonal portion of A and think Jacobi iteration. Then I showed that the operator-trigonometric meaning of this two-level scheme is

$$\gamma = \sin \phi(D^{-1}A) = \sin \phi(D^{-1/2}AD^{-1/2}).$$

For further discussion see [Gustafson (1997b)].

Then for presentation at the *Tenth International Conference on Domain Decomposition Methods* in Boulder in August 1997, I created a new trigonometric analysis of certain iterative domain decomposition methods and published them in the conference proceedings [Gustafson (1998a)]. However the proofs were too clumsy and so I deferred them to Sec. 3 of the later paper [Gustafson (2003a)].

Here is a typical result. These investigations and methods are rather technical and require the reader to be already versed in the domain decomposition technology so I shall just summarize here. Under the assumed conditions, the optimal convergence rate of the additive Schwarz domain decomposition algorithm is trigonometric and depends explicitly upon the operator angle $\phi(W^{-1/2}AW^{-1/2})$ according to

$$\rho\left(M^{\text{add SI}}_{\theta\,\text{optimal}}\right) = \sin\phi(W^{-1/2}AW^{-1/2}).$$

Notice how my trigonometric operator $\sin\phi(A)$ keeps appearing in these results, rather than $\cos\phi(A)$, the intuitive reason being that these iterative linear solvers live more in the spirit of relaxation, e.g., in the spirit of my norm curve $\|\epsilon A - I\|$, rather than in the variational spirit of my antieigenvalue functional $\mu(A)$.

See [Gustafson (2003a)] for further domain decomposition trigonometry. Also there I connect the recent SPAI (Sparse Approximate Inverse) algorithm developments to early ideas of mine in the operator trigonometry. I will not discuss those here.

4.6   Preconditioning and Condition Numbers

One of the more interesting new results I found in [Gustafson (2003a)] was the following rather general fact of importance for many preconditioning schemes. I state it as follows.

Lemma 4.2. (Much Preconditioning is Optimally Trigonometric). *Let A and B be SPD matrices on an inner product space X. Then the operator trigonometries of preconditioned BA, AB, $A^{1/2}BA^{1/2}$, $B^{1/2}AB^{1/2}$ (and many others) are all the same. The optimal (for many preconditioned $Ax = b \rightarrow BAx = Bb$ solvers) convergence radius is*

$$\rho^* = \min_{\epsilon} \|1 - \epsilon BA\|_C = \sin\phi(BA)$$

where C is any SPD matrix for which BA is SPD in the C inner product $\langle Cx, y\rangle$.

Then in the subsequent paper [Gustafson (2004a)] I obtained the characterization of all possible C inner products.

Lemma 4.3 (Triple Commute). *The class $\{C\}$ in which BA is SPD in C inner product is characterized equivalently by:*

(a) *$ABC = CBA$, i.e., A, B, C triple commute.*
(b) *CBA is SPD in the original inner product.*
(c) *BA is symmetric in the C inner product $\langle Cx, y\rangle$.*

The class $\{C\}$ forms a positive cone in the Hilbert space.

Similar results were found later in [Liesen and Saylor (2005)].

Finally, I would like to present a rather interesting related theoretical result about condition numbers in the Frobenius norm. I announced this result in 2003 at the NAPA Preconditioning Conference where I presented the paper [Gustafson (2003e)]. I sketched my analysis in [Gustafson (2006a)] and published it fully as a part of [Gustafson (2009b)]. I was following the paper by Chatelin and Gratton (2000) in which they obtained the following result for the absolute condition number $C(H)$ of an $n \times n$ real invertible matrix A written in right polar form $A = QH$ where $H = (A^*A)^{1/2}$ is the polar factor. $C(H)$ is defined to be the norm of the Fréchet derivative

$$C(H) = \|H'(A)\|_F = \limsup_{\substack{\Delta A \in R^{n \times n} \\ \|\Delta A\| \leqq \delta \\ \delta \to 0}} \frac{\|H(A) - H(A + \Delta A)\|_F}{\delta}.$$

Here the Frobenius norm is $\|A\|_F = \left[\sum_{i,j=1}^{n} |a_{ij}|^2 \right]^{1/2}$, also called the Hilbert–Schmidt norm. Their result was that

$$C(H) = \frac{\sqrt{2}(1 + \kappa^2(A))^{1/2}}{1 + \kappa(A)}$$

where $\kappa(A)$ is the usual condition number $\kappa = \sigma_1/\sigma_n$ ratio of largest to smallest singular value.

My result was that $C(H)$ has three other equivalent characterizations, written more conveniently here in terms of $C^2(H)$:

$$C^2(H) = 1 + \sin^2 \phi(H)$$

$$= \omega_b^{SOR}(H^2)$$

$$= \frac{\epsilon_m^2(H)}{\epsilon_m(H^2)}.$$

My goal was of course to render their Frobenius norm absolute condition number trigonometric rather than in terms of the old stretching ratio condition number κ. The above characterizations are easily obtained once you know my antieigenvalue theory.

More interesting and less easy was to show that in fact the dynamical meaning of $C(H)$ is a maximal turning action by the full operator A^*A. I refer the reader to the paper [Gustafson (2009b)] for the proof of that result.

Then I wanted to find a precise ΔA which actually gives the maximal turning seen as a dynamic trigonometric action. I used a singular value representation $\Delta A = UBV^T$ and found such a B, and hence a ΔA, where B is a kind of antieigenmatrix:

$$B = \begin{bmatrix} 0 & \cdots & 0 & \frac{\sigma_1}{(\sigma_1^2 + \sigma_n^2)^{1/2}} \\ \vdots & & & 0 \\ 0 & & & \vdots \\ \frac{\sigma_n}{(\sigma_1^2 + \sigma_n^2)^{1/2}} & 0 & \cdots & 0 \end{bmatrix}.$$

There are other related interesting new results the reader may find in [Gustafson (2009b)].

The point to take away is that all of $\kappa(A)$, $C(H)$, and $\sin \phi(H)$, are equally valid condition numbers. A large value of the conventional $\kappa(A)$ between 1 and ∞ means large relative stretchings along the major and minor principal axes. Large Frechét $C(H)$ between 1 and $\sqrt{2}$ means large derivative. My operator $\sin \phi(H)$ being large between 0 and 1 means large twisting. Which dynamics is more appropriate to your application?

Commentary

There are a number of interesting vignettes I would like to recount in this commentary about my return to my operator trigonometry in the 1990s. The first three come out surrounding the Dubrovnik meeting in June 1990. Recall that it was that meeting which gave me the first opportunity to present my instant rendering of the Kantorovich bounds for steepest descent and conjugate gradient convergence as fundamentally operator trigonometric.

First, as I flew in from Vienna, sitting beside me was a lovely coed of a joint American–Yugoslavian family, returning from UCLA. She whispered detailed predictions to me of a coming war, and why. I did not know if I should believe her. But the guards who rather roughly shook us down as we departed the airplane at the Dubrovnik airport added credence to her admonitions. This was several months before the war actually broke out. Where was the US intelligence community?

Second, I knew that my discovery of the fundamental direct trigonometric meaning of the Kantorovich convergence bounds would be of importance. So I intentionally put it into the proceedings of the Dubrovnik meeting. Then I waited. War broke out. I may have been the only meeting participant who truly was concerned as to whether the proceedings would in fact appear. Somehow, the editors (Časlav Stanojević and Olga Hadžić) did manage to publish the proceedings, and hence my contribution [Gustafson (1991a)].

Third, I submitted the paper [Gustafson (1994a)] in 1993, a paper essentially invited by the organizers of a 1992 numerical range conference to which I had been unable to go. In the manuscript I cited my paper in the Yugoslavian meeting proceedings and its result. Certainly those proceedings were not very accessible to the general public. A couple of months later I received a very confidential fax from a rather secretive-sounding commercial agency, wanting a copy of my Yugoslavian paper, citing it as such, and offering to pay me a sum for the paper. I was to send the copy of my paper to the commercial firm's address. They said they were getting the paper from me for some client who was to remain unnamed. This is probably the strangest "preprint request" I have ever received. In any case, I did not comply.

When I first presented my application of the operator trigonometry to convergence rates for iterative linear solvers at the Colorado Conference on Iterative Methods at Breckenridge, Colorado, in April 1994, a colleague from Texas who knew and appreciated some of my earlier work on finite element methods quietly took me aside. He then gave me his opinion that no one there would understand nor appreciate this noncommutative operator trigonometry as it applied to their matrices. Indeed, my full paper [Gustafson (1994c)] was rejected by a SIAM journal from the conference. I then published my results in [Gustafson (1996b; 1997b)]. At the 1996 conference in the Netherlands at which I presented the first of these papers, a very well known computational linear algebraist from Germany after my presentation asked me "but can you produce a better algorithm?" Someone else there later pointed out to me that my questioner, although a great authority and author of several excellent books on computational linear algebra, had never himself "produced a better algorithm." Nevertheless, I do respect that viewpoint in that field of scientific endeavor: producing a better algorithm is the name of the game. It is more important than a better theory, which in my opinion, I have indeed produced.

4.7   Exercises

1. (Historical). Look up and contrast the work of D. M. Young and my Ph.D. student D. P. Young.

2. As an exercise in operator trigonometry, derive the identity

$$\sin \phi(A^{1/2}) = \frac{\sin \phi(A)}{1 + \cos \phi(A)}$$

for an SPD $n \times n$ matrix A.

3. Run the conjugate gradient scheme on the Model Problem

$$\begin{bmatrix} 4 & -1 & -1 & 0 \\ -1 & 4 & 0 & -1 \\ -1 & 0 & 4 & -1 \\ 0 & -1 & -1 & 4 \end{bmatrix} \begin{bmatrix} x_1 \\ x_2 \\ x_3 \\ x_4 \end{bmatrix} = \begin{bmatrix} 1 \\ 2 \\ 0 \\ 0 \end{bmatrix}$$

starting from $x_0 = [0,0,0,0]^T$. Then do it starting from the first antieigenvector $x_+ = [0.683, 0.183, 0.183, 0.683]^T$. Track the turning angles and other operator trigonometric entities through the iterations.

4. Calculate the first antieigenvectors $x_\pm$ for the general discretized matrix A_h of the Model Problem, and for the 4×4 coarse mesh example.

5. It is quite straightforward to demonstrate that absolute Frobenius condition number $C(H)$ is in the relation

$$C^2(H) = 1 + \sin^2 \phi(H)$$

with my operator trigonometry. Do so.

6. The Triple Commute Lemma (Lemma 4.2) of Sec. 4.6 is nice and clean but one sees more if one tries to actually compute the whole class $\{C\}$ of available inner products. Do that for

$$A = \begin{bmatrix} 7 & 2 \\ 2 & 4 \end{bmatrix} \quad \text{and} \quad B = \begin{bmatrix} 1/7 & 0 \\ 0 & 1/4 \end{bmatrix}.$$

Chapter 5

Applications in Wavelets, Control, Scattering

Perspective

In the period roughly 1980–2002 I worked in mathematical physics with colleagues in Brussels at the Solvay Institute in which I became an Honorary Member. In particular, we developed a Time-operator theory for statistical physics and stochastic processes of Kolmogorov type. Those stochastic processes enjoy multiresolution structures quite similar to those that became popular later in wavelet theory. At the 1985 Alfred Haar Memorial Conference in Budapest, I specifically linked our theory of dynamical unstable coarse-grained irreversible processes and their internal Time-operators to general Haar systems. See [Gustafson (1987a)]. So as the wavelet theory burst upon the scene in the later 1980s, I was quite aware of its developments and could see the strong overlaps with our theory. But the motivations were totally different. The great wavelet explosion came in my opinion from three communities: the coherent state physicists, the signal processing engineers, and the Fourier and group-representing mathematicians. We were a fourth group, the stochastic physicists. I have tried to make all this clear in a couple of recent invited survey papers [Gustafson (1999g; 2007c)].

In particular in 1991–1992 our use of the Foias–Nagy–Halmos dilation theory made it clear to us that wavelet subspaces were just wandering subspaces and that in those terms we could obtain an explicit Time operator for all wavelet multiresolutions. I wrote a rough paper draft but we were all busy with other things and the paper never got published until years later [Gustafson (2000e)]. Section 5.1 of this chapter recounts those developments.

The preceding chapter treated my application of the operator trigonometry to numerical linear algebra. When I was invited to the 1997 IMACS Iterative Methods Symposium in Jackson Hole, Wyoming, I decided to connect my operator trigonometry of iterative linear solvers as discussed here in Chapter 4, to wavelet frames, see [Gustafson (1998b)]. Those results are presented here in Sec. 5.2. In particular I showed that the wavelet reconstruction algorithm is just Richardson iteration. To this day I am sometimes surprised to read papers in the wavelet literature that seem unaware of this fact. Further details are given in Sec. 5.3.

My first draft of this chapter then proceeded to three minor applications of my antieigenvalue theory: to neural networks, control, and scattering theories, in that order. Sections 5.5 and 5.6 on control and scattering remain as before and are admittedly in my opinion relatively inconsequential. But Sec. 5.4 now introduces instead a new Basis Trigonometry which I propose here for the first time. It is an outgrowth of the Time-operator theory for wavelets presented in Sec. 5.1, but will go far beyond that context.

5.1   The Time-Operator of Wavelets

Wavelets as a generalization of Fourier analysis and special function theory to a new breed of orthonormal bases is a wonderful addition to mathematics. As I described above, it was very natural to see rather early on that the five fundamental properties of a wavelet multiresolution analysis (MRA) were very nearly the same as five properties of Kolmogorov dynamical systems $(\Gamma, \beta\mu, s^n)$ for which we had already created a Time-operator.

My coworker Ioannis Antoniou in Brussels and I in Colorado were very busy with other duties and other research and we did not publish our Time-operator paper until much later [see Gustafson (2000e)]. However, I presented our results in my 1995 Japan lectures [see Gustafson (1996a; 1997d)]. In fact I gave in Part II of that book a rather extensive presentation of the four-way connection we established between Kolmogorov systems, wavelet multiresolution analysis, bilateral shifts with wandering cyclic subspaces, and regular stationary stochastic processes. It would be redundant to repeat all that here. Therefore let me restate Theorem 2.1.3 found in p. 83 of that book [Gustafson (1997d)].

Theorem 5.1 (Wavelet Systems and Kolmogorov Systems).

(a) *Wavelet multiresolution analyses and Kolmogorov systems possess the same basic five properties*:

(1) *a nesting structure,*
(2) *a trivial intersection,*
(3) *a full union,*
(4) *a refinement process,*
(5) *a regularity property.*

The principal distinction appears in the refinement properties (4). *There the wavelet MRA is generally constrained to an underlying scaling (viz., dilation, stretching) group, whereas the K-system is constrained to an underlying measure-preserving group.*

(b) *One may define any multiresolution analysis on $\mathcal{L}_{\mathbb{R}}^2$ as a bilateral shift of scalings on $\mathcal{L}_{\mathbb{R}}^2$ with countable infinite multiplicity such that the wandering generating subspace is a cyclic representation space with respect to the group of discrete translation. The cyclic vectors are just the wavelets. This statement amounts to an operator theoretic characterization of the scaling transformations of wavelets.*

The five properties referred to in Theorem 5.1 are interwoven throughout Chapter 2 of Part II of the book [Gustafson (1997d)] and I refer the reader to those pages there. In short, Property (5) of the wavelet multiresolution analysis is an irreducibility requirement common to all theories related to regular representations in group theory. It is shared by both wavelet MRA and K-systems. The space H_0 of the MRA corresponds to the projection P_0 of the coarse-graining induced by the generating partition ξ_0 of the K-system. We found the wandering subspace characterization ourselves but I have learned since then that others also found it because it is so natural once one gets into such nested mathematical subspace structures connected by translation or dilation groups.

Because we already had a Time-operator from our work in statistical physics, it was a natural step (Theorem 2.3.1 of [Gustafson (1997d)]) to wavelets.

Theorem 5.2 (Time-operator of Wavelets). *Any   wavelet   multiresolution analysis defines a Time-operator. The multiresolution approximation projections P_m are the spectral projections of the Time-operator T. The wavelets $\psi_{m,n}$ are the age eigenstates of the Time-operator. Age m means detailed resolution at the stage m. The action of the Time-operator in terms of wavelets is*

$$Tf(x) = \sum_{m\in\mathbb{Z}} \sum_{n\in\mathbb{Z}} m\langle\psi_{m,n}, f\rangle\psi_{m,n}(x).$$

Please note that I have changed the wavelet indices here from $\psi_{n,\alpha}$ as we originally had them e.g., as in [Gustafson (1997d)] to $\psi_{m,n}$ to agree with the more conventional wavelet literature notation, e.g., see [Daubechies (1992)]. If one goes to my later papers [Gustafson (1999g; 2007c)] where I better explained our approach to wavelets, you might also want to make that change: just replace our n, α with m, n. We were working quite independently and came from notations we had used for Kolmogorov dynamical systems.

Later it became clear that our approach to wavelets had as a consequence been different from the traditional one. We dealt with the shift V (the dilation) first, via its wandering subspaces W_m. This allowed us to get our Time-operator. The conventional wavelet theory does not do that. In fact in a series of recent papers by Levan and Kubrusly (2003; 2004; and later) have followed our approach to develop an alternate theory of wavelet MRA based upon our viewpoint that the MRA is seen to approximate arbitrary $f(x)$ in $\mathcal{L}^2(\mathbb{R})$ as the sum of its layers of details over all time shifts rather than as the sum of its layers of details over all scales.

As this monograph is about my antieigenvalue analysis and operator trigonometry, and not about Kolmogorov dynamical systems, I refer the reader to the book [Gustafson (1997d)] and our paper [Gustafson (2000e)] for the details of proof about our Time-operator. A key tool is the canonical commutation relations of quantum mechanics. However, let me mention two more results here before passing to the next section. The first is our reinterpretation of Haar's basis. The second is new and will be further developed in Sec. 5.4.

One of the goals of A. Haar's famous 1910 dissertation was to produce a complete orthonormal basis that was not derivable from any second-order Sturm–Liouville selfadjoint differential operator. A consequence of our Time-operator theory is that Haar's basis is now seen naturally as an eigenbasis of a first-order selfadjoint differential operator. Perhaps I should be more careful and state the result as follows. See [Gustafson (1998h)].

Theorem 5.3 (Haar Basis Differential Operator). *The Time-operator of statistical physics is the natural operator for which the Haar orthonormal system is its eigenbasis. As a (position) multiplication operator, it is naturally equivalent under appropriate transform to a first-order (momentum) differential operator.*

Finally, as I was writing this section, and thinking about our Time-operator for the Haar system, and indeed our Time-operator for any wavelet multiresolution system, it dawned on me that I could now extend my operator trigonometry to any wavelet basis generated by any multiresolution system. This will be a new trigonometry for wavelet bases. I will develop this idea further in a preliminary way in Sec. 5.4. But let me give an example here to get the ball rolling.

To do so, I borrow from the book [Daubechies (1992), pp. 10–14]. In particular in Fig. 1.5 on p. 11 she draws the two Haar wavelets $\psi_{1,1}$ and $\psi_{3,0}$. I will use those for the example here. Generally, the wavelets $\psi_{m,n}(x)$ are generated from $\psi(x)$ according to

$$\psi_{m,n}(x) = 2^{-m/2}\psi(2^{-m} - n).$$

Two Haar wavelets of the same scale (same m) never overlap. The support of $\psi_{m,n}$ is $[2^m n, 2^m(n+1))$. Thus $\psi_{1,1}$ oscillates once about $x = 3$ and has support $[2, 4)$. Similarly $psi_{3,0}$ oscillates once about $x = 4$ and has support $[0, 8)$. Because $\psi_{1,1}$ and $\psi_{3,0}$ are eigenbasis elements for our Time-operator, I may inquire into the corresponding wavelet antieigenvectors that they generate. Such are, according to my combinatorial antieigenvector viewpoint put forth in Chapter 3, defined according to

$$\psi_{\pm} = \left(\frac{m_{\text{older}}}{m_{\text{older}} + m_{\text{younger}}}\right)^{1/2}\psi_{\text{younger}} \pm \left(\frac{m_{\text{younger}}}{m_{\text{older}} + m_{\text{younger}}}\right)^{1/2}\psi_{\text{older}}$$

$$= \left(\frac{3}{4}\right)^{1/2}\psi_{1,1} \pm \left(\frac{1}{4}\right)^{1/2}\psi_{3,0}$$

$$= \frac{1}{2}[\sqrt{3}\psi_{1,1} \pm \psi_{3,0}].$$

I leave it to Exercise 1 at the end of this chapter to draw these and work out a few further details. In particular these two antieigenvectors $\psi_\pm$ produce an age turning angle of 30°.

So now we have antieigenvectors for any wavelet system. If that is generated by a multiresolution analysis, these antieigenvectors may be regarded as turning vectors for the corresponding Time-operator. My observation goes further: you do not even need a multiresolution analysis. You do not even need an operator. Given any orthonormal system, you may now form its two-component antieigenvectors and generate a turning trigonometry from them. If that system is an eigensystem from some particular operator, I would suggest that you first use the weights from my antieigenvalue theory. But there are other interesting weight coefficients as we shall see later in this book.

5.2 Frame Operator Trigonometry

Because wavelet theory seen within the larger (and older) context of frames becomes, in essence, variational, and hence closer in spirit to my antieigenvalue theory, and because I was quite familiar with numerical linear algebra relaxation methods as described in Chapter 4, I treated some numerical linear algebra aspects of the wavelet theory in the conference paper [Gustafson (1998b)]. In this section I will give the associated frame operators S some operator trigonometry. In the next section I will show how the wavelet reconstruction algorithm is Richardson iteration and hence automatically has the operator trigonometry I already worked out (see Theorem 4.1) for it in the previous chapter.

Frames go back to the fundamental paper of [Duffin and Schaeffer (1952)]. An abstract frame is any sequence of nonzero vectors $\{x_n\}$ such that there exist positive constants A and B such that for any x in that Hilbert space $\mathcal{H}$,

$$A\|x\|^2 \leqq \sum_n |\langle x, x_n \rangle|^2 \leqq B\|x\|^2.$$

A frame $\{x_n\}$ is a complete basis but is also allowed to be overcomplete, i.e., not linearly independent. Nonetheless any frame uniquely determines a corresponding SPD frame operator S defined by

$$Sx = \sum_n \langle x, x_n \rangle x_n.$$

One hopes the frame bounds, A and B, are reasonably sharp, but in any case we have the variational context

$$0 < A \overset{\leq}{=} m \equiv \|S^{-1}\|^{-1} \overset{\leq}{=} \frac{\langle Sx, x \rangle}{\langle x, x \rangle} \overset{\leq}{=} \|S\| \equiv M \overset{\leq}{=} B.$$

The frame is called tight if $A = B$ and called exact if you cannot remove any of the x_n. Orthonormal bases are exact tight frames; and conversely. Orthonormal wavelets $\{\psi_{nm}\}$ obtained by discrete dilation and translation of a single waveform ψ are tight exact frames with $A = B = 1$.

I observed in [Gustafson (1998b)] that we may immediately define an exact or approximate frame operator maximal turning angle $\phi(S)$ according to

$$\cos\phi(S) = \frac{2\sqrt{AB}}{A + B} \quad \text{or} \quad \sin\phi(S) = \frac{B - A}{B + A}.$$

By now that should be rather obvious to the reader of this book. Likewise it should be clear that if the frame is tight, the frame operator S has angle $\phi(S) = 0$. So tightening a frame corresponds to reducing the turning angle of S.

For example, in [Daubechies (1992), p. 71] for wavelet frames $\{\psi_{m,n}\}$ obtained from a wavelet function ψ with dilation parameter a_0 and translation parameter b_0, i.e., $\psi_{m,n}(x) = a_0^{-m/2}\psi(a_0^{-m}x - nb_0)$, the estimate

$$A \overset{\leq}{=} \frac{2\pi}{b_0} \sum_m |\hat{\psi}(a_0^m \xi)|^2 \overset{\leq}{=} B$$

links the frame $\psi_{m,n}$ to the width of the wavelet ψ in the frequency domain. We see that we now interpret any such discussion in terms of the frame angle $\psi(S)$. For example, by using a technique of multivoices per octave, one can obtain (in our perspective) a frame angle $\phi(S)$ close to zero for the Mexican hat function. See [Daubechies (1992)] Table 3.1 on p. 77. On the other hand, we see that the size of the frame angle $\phi(S)$ departs radically from small values as wavelet translation parameter b_0 is taken too large. Similarly for "filter-based" example (pp. 78–80) and dilation parameter $a_0 = 2$, we see that $b_0 = 0.5$ and $b_0 = 1$, respectively, produce frame operator angles

$$\frac{B - A}{B + A} = \frac{2.66717 - 2.33854}{2.66717 + 2.33854} = 0.06565 \Rightarrow \phi(S) = 3.7642°,$$

$$\frac{B - A}{B + A} = \frac{1.77107 - 0.73178}{1.77107 + 0.73178} = 0.41524 \Rightarrow \phi(S) = 24.535°.$$

Looking at the Table 3.3 on its p. 87, we see that when $\omega_0 t_0/2\pi$ approaches 1, B/A becomes very large, and to us that corresponds to the frame operator S having a turning angle approaching 90°. When $\omega_0 t_0/2\pi = 1/N$, one can compute exact frame bounds A and B. In our terms, that means we know the frame operator angle exactly.

Turning to Daubechies' fine paper [Daubechies (1990)], we see for example from Table IV there for $g(x) = \exp(-|x|)$ we may know the exact frame bounds and thus the exact frame angles $\phi(S)$. But when we do not know the upper frame bound B exactly, it behooves us to have reasonably sharp lower bounds for B if we want to have good upper bounds for our frame operator turning angle. That is,

$$\sin \phi(S) \overset{\le}{=} \frac{(\text{Upper bound for } B) - A}{(\text{Lower bound for } B) + A}.$$

Intuitively, a sharp lower bound for B means an estimate quite near to the highest eigenvalues of S, but below it.

The nice thing about frames as contrasted to wavelets is that we actually have an associated linear operator. Moreover, trigonometrically speaking, they go beyond orthonormal wavelet bases for which the frame operator S has turning angle $\phi(S) = 0$. I have not pursued this frame operator trigonometry any further, but clearly it becomes more interesting as the frame becomes less tight.

5.3 Wavelet Reconstruction is Trigonometric

Reconstruction of a function x from its wavelet frame coefficients $\{\langle x, x_n \rangle\}$ is an important task. The wavelet reconstruction algorithm is one of successive approximation. Regarding $Sx = \sum_n \langle x, x_n \rangle x_n = F^*Fx$ as representing the transform F of x to its frame coefficients $\{\langle x, x_n \rangle\} \in \ell_2$, followed by the adjoint transform F^* from ℓ_2 back to $\mathcal{H}$, leads to the reconstruction formula

$$x = \frac{2}{A+B} \sum_n \langle x, x_n \rangle x_n + Rx$$

where $R = I - 2/(A+B)S$. The error operator R satisfies

$$\|R\| \overset{\le}{=} \frac{B-A}{B+A} < 1.$$

See [Daubechies (1990; 1992)] for more information and details.

The reconstruction algorithm is the iteration (defect correction)

$$x^{N+1} = x^N + \frac{2}{A+B}(Sx - Sx^N).$$

We may place this wavelet reconstruction procedure within the operator trigonometry context as follows. For simplicity we assume that the frame bounds A and B are exact, or at least very sharp.

Theorem 5.4. (Wavelet Reconstruction is Trigonometric) *The wavelet frame reconstruction algorithm is that of Richardson iteration*

$$x^{N+1} = x^N + \alpha(b - Sx^N)$$

with iteration matrix $R_\alpha = I - \alpha S$, $b = Sx$, where S is the given wavelet frame operator. The choice of $\alpha = 2/(A+B)$ is optimal and produces optimal Richardson convergence rate

$$\rho_{\text{opt}} = \sin S.$$

Moreover, $\alpha = 2/(A+B) = \epsilon_m$, the value of ϵ which produces $\sin \phi(S)$ in the operator trigonometry. When $A = m$ and $B = M$ are attained by the eigenvectors of S, then

$$\alpha = \frac{2}{A+B} = \frac{\langle Sx_+, x_+ \rangle}{\|Sx_+\|^2}$$

where x_+ is a first antieigenvector of the wavelet frame operator S. Similarly for x_-.

The theorem follows directly from our Theorem 4.1 on Richardson iteration given in Sec. 4.3.

When A and B are sharp frame bounds, then the error $\|R\| = (B-A)/(B+A)$ exactly. Even when A and B are only close approximate frame bounds, the theorem remains approximately valid and the wavelet frame reconstruction algorithm is nearly optimal Richardson iteration. That is the case because the operator trigonometry possesses considerable stability under small perturbations.

Daubechies states (1992), p. 61, intuitively, that from $A \overset{\leq}{=} S \equiv F^*F \overset{\leq}{=} B$, S is "close" to $(A+B)/2I$, and S^{-1} is "close" to $2/(A+B)$ I, to justify what we have now seen here to be an optimal Richardson iteration. Then by writing $(B-A)/(B+A) = r/(2+r)$, i.e., $r = (B-A)/A$, the wavelet frame reconstruction error Rx is seen to be small when r is small. That good intuition is sharpened and made exact here by replacing r by

$\sin\phi(S) = (B - A)/(A + B)$. In [Daubechies (1990)] she uses a different r, $r = B/A - I/(B/A + 1)$ both as a measure of how snug the frame is and also as the parameter ensuring convergence of the wavelet frame inversion algorithm. Of course now we may see from our theorem that this r is exactly the optimal convergence rate of the wavelet frame inversion algorithm, namely, $r = \sin\phi(S)$. Redundancy of a frame, i.e., how many "extra" vectors it contains, is measured roughly by the parameter $r = (A + B)/2$. Observe that this $r^{-1} = \epsilon_m$ = the optimal α of Richardson iteration. Finally we note, in analogy with our analysis of Chebyshev centers and midwidths in [Gustafson (1997b)] that

$$\sin\phi(S) = \sin\phi(S^{-1}) = \frac{(B - A)/2}{(B + A)/2} = \frac{\text{frame midwidth}}{\text{frame center}}.$$

From the above discussion, we conclude that especially for nontight frames, the frame operator trigonometry nicely incorporates all of the above mentioned measures of closeness, smallness, snugness, convergence, and redundancy. Within the trigonometry of the frame operator S, the quantity $\sin\phi(S)$ plays an especially important role.

5.4   New Basis Trigonometry

As shown in Sec. 5.1, introducing the Time-operator into the theory of wavelets conceptually led me to now also introduce antieigenvectors into wavelet theory. I wish to pursue those ideas a bit further here on a preliminary basis. Further investigation will have to follow elsewhere.

The example given in Sec. 5.1, namely the Haar basis antieigenvector

$$\psi_{\pm} = \frac{1}{2}[\sqrt{3}\psi_{1,1} \pm \psi_{3,0}]$$

was constructed from two convenient Haar wavelets. Bear in mind that there are 2^m wavelets at each mth level, so one can construct many others similar to the above $\psi_{\pm}$ from levels 1 and 3, using either of the two $\psi_{1,0}$ or $\psi_{1,1}$ and any of the eight from $m = 3$. For this example, all of them will resemble the one worked out in Exercise 1, the only difference being where their supports overlap and thereby increase or decrease each other there.

The Time-operator T as an unbounded accretive operator will have its overall operator $\cos\phi(T) = 0$, which in the past discouraged me from further interest for the original multiplicative perturbation application.

But now we are thinking more generally about any given eigensystem, or even any orthonormal basis. Using my antieigenvalue weights connotes a certain turning, or perhaps better thought of as a twisting caused by the counterweighting of the given two eigenvectors. One does not need a Time-operator, or even any operator at all, if the orthonormal basis does not have such. Then you do not have natural physical or mathematical guidance through the eigenvalues to tell you how to weigh any two basis elements. But we will see other interesting criteria for optimal weighting later in this book. And one could consider antieigenvectors of three, four, and more, components. What I have in mind will be a new concept of basis trigonometry — a trigonometry inspired by my operator trigonometry — but now with the emphasis on the basis itself. This is a natural concept within the context of wavelets, where the philosophy is orthonormal bases generated by, for example, a given cyclic vector, rather than by an operator. I will call this a *spectral twisting* theory, as a subject of interest in itself.

I have no idea what lies in the future of such a new spectral theory. Hopefully this book will lead others to look at it in other fields of application which will then guide its further development. In this section I stay with wavelets and play in a very preliminary way with a few examples.

First I again followed [Daubechies (1992), p. 115], where after introducing the Haar wavelets again, she brings in the Littlewood–Paley basis, generated from cyclic vector

$$\psi(x) = \frac{\sin 2\pi x - \sin \pi x}{\pi x}.$$

The corresponding $\psi_{m,n}(x) = 2^{-m/2}\psi(2^{-m}x - n)$ are an orthonormal basis with better frequency location than Haar wavelets, since their Fourier transforms are compactly supported.

No doubt it will be profitable to go eventually to Fourier transform space, the mathematician's proclivity in these matters. I or someone else can do that in the future. However, here I want to investigate the twisting trigonometry directly. It turns out to be quite intricate, deep, trigonometrically complicated, you may choose the descriptive phrase as you like. To me, it means there is a fabulously rich basis trigonometry of special functions waiting to be further developed.

Let us see how complicated the trigonometry that the Littlewood–Paley basis generates is. As this is a very preliminary investigation, sometimes I will not bother to make everything to be norm one.

First I mimicked my example of Sec. 5.1, using instead ψ_{11} and ψ_{30} from the Littlewood–Paley basis. By my algebra these are

$$\psi_{11}(x) = \frac{1}{\sqrt{2}} \frac{[\sin 2\pi(2^{-1}x - 1) - \sin \pi(2^{-1}x - 1)]}{\pi(2^{-1}x - 1)}$$

$$= \frac{2}{\sqrt{2}} \frac{[\sin \pi x + \sin (\pi x/2)]}{\pi(x - 2)}$$

and

$$\psi_{3,0}(x) = 2^{-3/2} \frac{[\sin 2\pi(2^{-3}x) - \sin \pi(2^{-3}x)]}{\pi(2^{-3}x)}$$

$$= \frac{4}{\sqrt{2}} \frac{[\sin (\pi x/4) - \sin (\pi x/8)]}{\pi x}.$$

From these I formed the corresponding antieigenvectors (I will keep using that term here, rather than twisting vector)

$$\psi_{\pm} = \left(\frac{3}{4}\right)^{1/2} \psi_{11} \pm \left(\frac{1}{4}\right)^{1/2} \psi_{30}$$

and came to a rather complicated trigonometric function. I will let the reader grapple with that in Exercise 5.

So I backed off to ψ_{10} and ψ_{30}, from which came the antieigenvectors

$$\psi_{\pm} = \left(\frac{3}{4}\right)^{1/2} \psi_{10} \pm \left(\frac{1}{4}\right)^{1/2} \psi_{30}$$

$$= \frac{1}{\sqrt{2}} \cdot \frac{1}{\pi x} \left[\sqrt{3}\sin \pi x - \sqrt{3}\sin\left(\frac{\pi x}{2}\right) \pm 2\left(\sin\left(\frac{\pi x}{4}\right) - \sin\left(\frac{\pi x}{8}\right)\right)\right].$$

Rather nice (complicated) trigonometry is involved. Using ψ_{10} and ψ_{20} gave after some algebra the antieigenvector (I just did ψ_{+} for this one)

$$\psi_{+} = \left(\frac{2}{3}\right)^{1/2} \psi_{10} + \left(\frac{1}{3}\right)^{1/2} \psi_{20}$$

$$= \frac{1}{\sqrt{6}} \cdot \frac{1}{\pi x} \left\{\left(2 + \frac{1}{\sqrt{2}}\right) \sin (\pi x) - \left(2 + \frac{1}{\sqrt{2}}\right) \sin\left(\frac{\pi x}{2}\right)\right\}$$

$$= \frac{(2\sqrt{2} + 1)}{4} \psi\left(\frac{x}{2}\right).$$

This is a bit more pleasing, if my algebra holds up. Finally I wondered about two wavelets from the same level. Since at this preliminary stage of my investigation I am still using the Time-operator antieigenvector weighting scheme, for ψ_{10} and ψ_{11} the $m = 1$ weights are the same and I arrived at (dropping the $1 + 1 = 2$ weight denominator)

$$\psi_{10} + \psi_{11} = \frac{1}{\sqrt{2}} \left[\frac{\sin(\pi x) - \sin(\pi x/2)}{\pi x} + \frac{\sin(\pi x) + \sin(\pi x/2)}{\pi x - 2\pi} \right]$$

$$= \frac{\sqrt{2}}{\pi x(x - 2)} \left[(x - 1) \sin \pi x + 2 \sin \left(\frac{\pi x}{2} \right) \right].$$

It does not matter if my algebra is all okay, and I leave the checking to Exercise 5. What is evident is that we are generating a lot of interesting new algebraic-trigonometric functions. One could go to symbolic software to do much more with these.

At this point, because I need to finish the writing of the rest of this book, I was about to consign any further investigation of this new basis trigonometry to the future. However, I decided to go back to the most classic wavelet I know, namely, the Fourier sine eigenbasis of the fundamental Sturm–Liouville boundary value problem

$$\begin{cases} -u''(x) = \lambda u(x), & 0 < x < \pi, \\ u(0) = u(\pi) = 0, \end{cases}$$

whose eigenvalues are $\lambda_n = 1, 4, 9, \ldots, n^2, \ldots$ with eigenfunctions $\psi_n(x) = \sin nx$, $n = 1, 2, 3 \ldots$. What is its new basis trigonometry? Staying with $\psi_1(x)$ and $\psi_3(x)$ in the spirit of the above wavelet examples, we have the corresponding antieigenvector (I start with ψ_- here, with a reason)

$$\psi_- = \left(\frac{\lambda_3}{\lambda_1 + \lambda_3} \right)^{1/2} \psi_1 - \left(\frac{\lambda_1}{\lambda_1 + \lambda_3} \right)^{1/2} \psi_3$$

$$= \left(\frac{9}{10} \right)^{1/2} \sin x - \left(\frac{1}{10} \right)^{1/2} \sin 3x$$

$$= \text{const} \cdot \left[\frac{3}{4} \sin x - \frac{1}{4} \sin 3x \right]$$

$$= \text{const} \cdot \sin^3(x).$$

The inessential constant is $4/\sqrt{10}$, and also to get ψ_1 and ψ_3 originally of $L^2(0, \pi)$ norm 1, I would need to divide by their norm ($\int_0^\pi \sin^2 nx dx)^{1/2} = (\pi/2)^{1/2}$ which fortunately is the same for all n. But such is not the point here, since any nonzero scalar multiple of an antieigenvector remains an antieigenvector. Just note for the moment that I have not normalized things here.

What pleasantly surprised me in this classical Rayleigh eigenvalue problem which is so basic to so much of science is that the antieigenvector

$$\psi_-^{1,3} = \sin^3(x) = \frac{3}{4} \sin x - \frac{1}{4} \sin 3x$$

constitutes a well known identity used to reduce $\sin^3(x)$ data to more manageable linear Fourier-compatible data. See my PDE book [Gustafson (1999f)], p. 355. This coincidence sufficiently intrigued me to work out the other antieigenvector

$$\psi_+^{1,3} = \frac{3}{4} \sin x + \frac{1}{4} \sin 3x$$

which turned out to be after some elementary trigonometry,

$$\psi_+^{1,3} = -\sin^3 x + \frac{3}{2} \sin x.$$

Because the turning angle here of our one-dimensional Laplacian differential operator with Dirichlet boundary conditions when restricted to the span of its first and third eigenvectors is given by

$$\sin \phi_{1,3}(A) = \frac{9-1}{9+1} = \frac{8}{10},$$

we find a turning angle $\phi_{1,3}(A) = 53.13010235°$.

Thus we have arrived at a new basis trigonometry even for the classical Rayleigh eigenvalue problem for the Laplacian operator. We no longer are constrained to eliminate from our interest unbounded SPD operators just because their overall first antieigenvalue is $\mu_1 = \cos \phi(A) = 0$.

As I pointed out in my PDE book [Gustafson (1999f), p. 355, footnote], one can arrive at identities such as

$$\sin^3(x) = \frac{3}{4} \sin x - \frac{1}{4} \sin 3x$$

from De Moivre's theorem

$$[r \cos \theta + i \sin \theta]^n = r^n[\cos n\theta + i \sin n\theta].$$

Thus from complex polar representation we have

$$e^{i3\theta} = (\cos\theta + i\sin\theta)^3 \equiv \cos 3\theta + i\sin 3\theta$$
$$= (4\cos^3\theta - 3\cos\theta) + i(3\sin\theta - 4\sin^4\theta)$$

whose imaginary part gives the above identity. Similarly

$$\cos^3(x) = \frac{3}{4}\cos x + \frac{1}{4}\cos 3x$$

and that would be the x_+ antieigenvector of the Rayleigh eigenvalue problem for an interval centered about $x = 0$. In the same way we have from De Moivre's theorem

$$\sin^5\theta = \frac{10}{16}\sin\theta - \frac{5}{16}\sin 3\theta + \frac{1}{16}\sin 5\theta$$

and

$$\cos^5\theta = \frac{10}{16}\cos\theta + \frac{5}{16}\cos 3\theta + \frac{1}{16}\sin 5\theta.$$

Such relations might indicate that my proposed new Basis Trigonometry should include also three-component antieigenvectors, as I also suggested in Exercise 6 of Chapter 3. Then there will be a further generalized n-component twisting trigonometry seen by going on to Pascal's triangle (to save time from doing De Moivre's expansion each time)

$$
\begin{array}{c}
1\\
1\ \ 1\\
1\ \ 2\ \ 1\\
1\ \ 3\ \ 3\ \ 1\\
1\ \ 4\ \ 6\ \ 4\ \ 1\\
1\ \ 5\ \ 10\ \ 10\ \ 5\ \ 1\\
1\ \ 6\ \ 15\ \ 20\ \ 15\ \ 6\ \ 1\\
1\ \ 7\ \ 21\ \ 35\ \ 35\ \ 21\ \ 7\ \ 1
\end{array}
$$

from which

$$\sin^7(\theta) = \frac{35}{64}\sin\theta - \frac{21}{64}\sin 3\theta + \frac{7}{64}\sin 5\theta - \frac{1}{64}\sin 7\theta,$$
$$\cos^7(\theta) = \frac{35}{64}\cos\theta + \frac{21}{64}\cos 3\theta + \frac{7}{64}\cos 5\theta + \frac{1}{64}\cos 7\theta,$$

and so on. Notice that the cosines all carry positive coefficients so they would generalize my x_+ antieigenvectors. There could be interesting comparisons to the Pascal triangular algorithm for B-spline wavelets discussed in Chui (1992).

Looking into this general new Basis Trigonometry in this section has brought out an interesting cosmetic improvement to my original convention for the antieigenvectors. Originally and usually I have placed the $\pm$ signs on the first component of the antieigenvectors:

$$x_\pm = \pm \left(\frac{\lambda_n}{\lambda_1 + \lambda_n} \right)^{1/2} x_1 + \left(\frac{\lambda_1}{\lambda_1 + \lambda_n} \right)^{1/2} x_n.$$

However I notice that in this chapter for convenience I seem to have slipped into placing the $\pm$ signs on the second component of the now seen more generally, antieigenvectors, e.g.,

$$\psi_\pm = \left(\frac{\lambda_n}{\lambda_1 + \lambda_n} \right)^{1/2} \psi_1 \pm \left(\frac{\lambda_1}{\lambda_1 + \lambda_n} \right)^{1/2} \psi_n.$$

In Sec. 3.2 I commented that you can place the $\pm$ sign on either the first or second component of the antieigenvectors, and that is indeed so because arbitrary scalar multiples of an antieigenvector retains the same turning angle.

Then in Sec. 3.4 I noted the result from [Gustafson (2000d)], that the antieigenvector pair enjoys an angular relationship

$$\langle x_+, x_- \rangle = - \sin \phi(A)$$

and thus they have intra-angle $\phi(A) + \pi/2$.

Having blithely neglected to normalize the trigonometric basis vectors I played with in this chapter, and you will want them to have the same norm unless you have other scaling or weighting desires, I decided I should just normalize in the last Rayleigh eigenvalue problem as a check on the new trigonometry found there. A result will be that it is probably better in most circumstances to put the $\pm$ sign on the second component of the antieigenvector.

Thus, normalized by their $L^2(0, \pi)$ norm, we have

$$\psi_- = \frac{4}{\sqrt{10}} \left[\frac{3}{4} \sin x - \frac{1}{4} \sin 3x \right] \Big/ \left(\frac{\pi}{2} \right)^{1/2}$$

$$= \frac{1}{\sqrt{5\pi}} [3 \sin x - \sin 3x],$$

and similarly

$$\psi_+ = \frac{1}{\sqrt{5\pi}}[3\sin x + \sin 3x].$$

Then if we look at the antieigenvector pair angle we find

$$\langle\psi_+, \psi_-\rangle = \frac{1}{5\pi}\int_0^\pi (3\sin x + \sin 3x)(3\sin x - \sin 3x)dx$$

$$= \frac{1}{5\pi}\left[9\left(\int_0^\pi \sin^2 x\,dx\right) - \left(\int_0^\pi \sin^2 3x\,dx\right)\right]$$

$$= \frac{8}{5\pi}\left(\frac{\pi}{2}\right) = \frac{4}{5}.$$

Thus with the $\pm$ placed on the second component, we have the more pleasing result:

Lemma 5.1 (Antieigenvector Pair Angle). *With the above $\pm$ convention, given any normalized antieigenvector combinatorial pair $x_\pm$, then the cosine of the angle between them is $\sin\phi(A)$:*

$$\langle x_+, x_-\rangle = \sin\phi(A).$$

The proof is the same as in [Gustafson (2000d)] with the signs changed. Note that now we have the obvious fact that $x_+ + x_-$ is (twice the weighting constant times) the first eigenvector or more generally first basis vector ψ_1. This is exhibited clearly in my Rayleigh problem example above, where the (unnormalized) antieigenvectors gave

$$\psi_+ + \psi_- = \frac{3}{2}\sin x.$$

5.5   Trigonometry of Lyapunov Stability

I knew at the beginning that operator trigonometry could have interesting interactions with control theory and this is mentioned in [Gustafson (1972a)]. However, by the time I went back to look at the situation almost 30 years later, I found that the operator theorists within control theory had greatly advanced the mathematical strength of the subject. So I was extremely brief in [Gustafson (1998d)] and I will be extremely brief here.

My idea in 1972 concerned the classic Lyapunov sufficient stability condition for the ODE (ordinary differential equation) system

$$\begin{cases} \dot{x}(t) = Ax(t), & t > 0, \\ x(0) = x_0 & \text{given}, \end{cases}$$

namely, that there exist some SPD matrix B such that

$$\langle (A^T B + BA)x, x \rangle < 0$$

for all $x \neq 0$. However, $A^T B + BA = 2\,\text{Re}\,BA$, so my operator trigonometric criteria from my original 1968 semigroup and positive (or negative) operator product papers immediately give sufficient conditions for $\text{Re}\,BA$ to be dissipative.

For further (limited) discussion in my subsequent foray into possible useful trigonometries for application to control theory, see [Gustafson (1998d)]. My conclusion at that time was that my trigonometric sufficient conditions are overly restrictive in that environment, and far from necessary.

However, somewhat as a palliative for the self-criticism of the previous paragraph, the reader will find in Sec. 6.3 another setting coming from control theory in which future development of a corresponding operator trigonometry could be more interesting. That setting of canonical correlations is more naturally variational and therefore closer to my theory.

5.6 Multiplicative Perturbation and Irreversibility

Quantum scattering theory as studied mathematically concerns unitary group exponentiations of a time-dependent Schrödinger equation:

$$\frac{du}{dt} = iHu$$

to the solution given by a one-parameter unitary group

$$u(t) = e^{iHt}u_0 \quad -\infty < t < \infty$$

applied to a prepared initial state u_0. In spite of huge volumes of literature over one hundred years, exactly what happens to the evolution during a measurement is still shrouded in controversy. I have worked on such questions over the years and was a pioneer in trying to create mathematical models such as the counter-problem, and in dealing with issues such as the quantum Zeno paradox. See for example the citations in [Gustafson (2006a)].

Physical operators are always subject to small disturbances: decoherence due to interaction with a noisy environment or other systems, etc., and under convenient assumptions the reversible unitary evolution is imagined to become an irreversible semigroup evolution. There are many serious and difficult issues about quantum measurement that are too extensive to go into here.

However, because historically my antieigenvalue theory was born from a context of scattering perturbation theory with its unitary group and semigroup evolution systems, and because the literature contains much treatment of additive perturbations but not much on multiplicative perturbations, I thought to go back to see if I could apply my operator trigonometry in any useful way. This I did in the short paper [Gustafson (1998c)]. Because my operator trigonometry theory works best for strongly accretive or selfadjoint positive operators, my results on this one foray were rather limited.

For simplicity consider bounded positive definite operators H and B on a Hilbert space and let $A = iH$. The $e^{At} = e^{iHt}$ is a unitary evolution with spectrum

$$\sigma(e^{itH}) = e^{it\sigma(H)} = \cos(t\sigma(H)) + i\sin(t\sigma(H))$$

consisting only of phase and no decay. B is to represent a multiplicative disturbance. Many physicists would just assume B commutes with H so that BH remains selfadjoint, but from the beginning I have not wanted to do that. To me, B should represent some dynamic effect from outside of H's spectral family. Even so, e^{BAt} is still a semigroup evolution with spectrum

$$\sigma(e^{BAt}) = \cos(t\sigma(BH)) + i\sin(t\sigma(BH)).$$

That follows because $\sigma(BH)$ remains real. One can then show that we may write $BH = B_1 T_1 B_1^{-1}$ by using results by Wigner on weakly positive operators. For irreversibility we would like to know when BA is a dissipative semigroup generator.

Writing $BA = C + iD$ in terms of its real and imaginary parts. I was able to show the following. Admittedly now assuming that C commutes with H, and that H and D are positive definite, if

$$\sin \phi(D) \stackrel{\leq}{=} \cos \phi(H),$$

then $W_t = e^{BAt}$ is a dissipative contraction semigroup. In particular, there always exists such a small disturbance such that $W_t = e^{BAt}$ is a completely nonunitary contraction semigroup. Recall that a completely nonunitary contraction semigroup has no subspaces which remain invariant under the semigroup for all $t \stackrel{\geq}{=} 0$. So B has created a qualitative change in the nature of the dynamics of the quantum evolution.

We will come to much more interesting applications of my operator trigonometry to quantum mechanics in Chapter 7.

Commentary

For the historical record, that we independently happened upon the wandering subspace concept in 1991 for wavelets, is documented in a proposal for a small NATO grant which we used in the early 1990s to get back and forth between Boulder and Brussels. See http://www.auth.gr/chi/PROJECTSWaveletsKolmog.html. You will find wandering subspaces on p. 1 of the book Sz. Nagy and Foias (1970) on operator dilation theory, which we used extensively for our investigations of Kolmogorov systems. It was not until about ten years later that Nhan Levan told me of the early paper [Goodman, Lee, and Tang (1993)]. Because we did not know any better, we dealt with the dilation V as a shift with wandering subspaces before we dealt with the translations. That ignorance led to our new view of wavelets.

See the book [Gustafson (1997d)] and my survey papers [Gustafson (1999g; 2007c)] for some further lesser-known historical facts about wavelets. As a personal experience, I myself first heard the term "wavelet" used when I rough-necked on an oil prospecting boat in the Gulf of Mexico in summer 1956. A quote I like that you will find in my papers cited above and taken from a classic 1954 geophysics paper by Robinson is the following: "All these wavelets have the same minimum-delay shape."

As I mentioned in this chapter's *Perspective*, my first draft of this chapter planned a short Sec. 5.4 dealing with neural networks, a field in which I worked in for several years in connection with our Engineering Optical

Computing Center here. In 1992 my Ph.D. student Shelly Goggin and I looked at the famous Backpropagation learning algorithm for implementation into an optoelectronic neural network. Backpropagation essentially does a very slow steepest descent: very, very slow at the end. For that reason, Shelly and I created an angularly converging version which we called Angleprop. A brief summary of our simulations may be found in [Gustafson (2002c)]. But the results were inconclusive. And an algorithm we developed called CNLOR, connectionist nonlinear over-relaxation, converged much much faster, when it indeed converged. One seems to have that in neural learning algorithms: the turtle, like Backprop or our Angleprop, slow but robust, or the hare, like CNLOR or some Newton-type algorithms, fast but unreliable.

5.7 Exercises

1. Compute my example Haar wavelet antieigenvectors $\psi_\pm$ given at the end of Sec. 5.1 so that you can draw them both on the interval $[0, 8)$.

2. To really understand what makes wavelet theory work, you should do some reading on the canonical commutation relations of quantum physics and the Stone–Von Neumann Theorem of mathematical functional analysis. On the other hand, taking Fourier transforms of everything to then do all of your wavelet mathematics in frequency domain will obviate your need to understand many of the time-domain physical happenings.

3. Given a frame

$$A\|x\|^2 \lesseqgtr \sum_n |\langle x, x_n \rangle|^2 \lesseqgtr B\|x\|^2,$$

show that as an operator $A \lesseqgtr S \lesseqgtr B$, that as an operator $B^{-1} \lesseqgtr S^{-1} \lesseqgtr A^{-1}$, and that $\{S^{-1}x_n\}$ is a frame.

4. In terms of the m-accretive operator theory of Chapters 1 and 2, a frame operator S is A-accretive, although if A is not exact, then it is not exactly S's sharp lower bound $m_S \equiv \|S^{-1}\|^{-1}$. For what $\epsilon > 0$ can you guarantee that $\|\epsilon S - I\| < 1$? If A and B are not exact, what can you say about the location of the ϵ_m for which $\|\epsilon S - I\|$ attains its minimum $\|\epsilon_m S - I\| = \sin \phi(S)$?

5. Verify (or correct) my quick calculations of example basis antieigenvectors for the Littlewood–Paley wavelets. Do a more thorough job. Do
the same for other interesting wavelet families.

6. For the classic Rayleigh eigenvalue problem

$$\begin{cases} -u''(x) = \lambda u(x), & 0 < x < \pi, \\ u(0) = u(\pi) = 0, \end{cases}$$

now for the Rayleigh antieigenvalue problem, verify all the essentials of
the operator trigonometry as concerns the skew antieigenvector $\psi_+^{1,3}$ that
I sketched in Sec. 5.4. Go further and normalize everything to norm one
and then verify the $\sin \phi(A)$ value from the relation $\langle \psi_+, \psi_- \rangle = \sin \phi(A)$
using the L^2 inner product $\langle u, v \rangle = \int_0^\pi u(x)v(x)dx$. Show that the distance between the two antieigenvectors with the $\pm$ second component
convention is always given by

$$\|x_+ - x_-\|^2 = 2(1 - \sin \phi(A)).$$

Sketch that situation for Example 2 of Chapter 1.

Chapter 6

The Trigonometry of Matrix Statistics

Perspective

In [Gustafson (1999e)] I turned my attention to the application of my antieigenvalue theory and its operator trigonometry to matrix statistics. Although this breakthrough paper [Gustafson (1999e)] was initially summarily rejected by one referee, nonetheless the paper, upon its submission was in circulation and quickly attracted substantial interest from within the matrix statistics community. I succeeded in publishing all of [Gustafson (1999e)] within the paper [Gustafson (2002a)]. Later I was invited to the IWMS 2005 conference in Auckland, New Zealand, where both I and the most noted C. R. Rao spoke on my antieigenvalue theory as it applies to matrix statistics. As a result of those presentations, Eugene Seneta asked me to write the review paper [Gustafson (2006b)]. Recently I was asked to contribute the chapter [Gustafson (2010e)] for the *Oxford Handbook of Functional Data Analysis*.

The above-mentioned four papers drive the presentation of this chapter, although there is both less and more here. Other papers that may be consulted are [Gustafson (2001a; 2005a; 2006a; 2007b; 2008a; 2009b; 2010d)]. For other citations to work of others, see the *Commentary* at the end of this chapter.

6.1 Statistical Efficiency

In this section I take up the theory of statistical efficiency. As a matter of general interest, I gave some history of statistical efficiency in [Gustafson (1999e; 2002a)], and I expanded that history in [Gustafson (2005a)]. There is also a very good historical account in [Chu, Isotalo, Puntanen, and Styan (2005)].

The following was shown in [Gustafson (1999e)]. Consider the general linear model

$$y = X\beta + e$$

where y is an n-vector composed of n random samplings of a random variable Y, X is an $n \times p$ matrix usually called the design or model matrix, β is a p-vector composed of p unknown nonrandom parameters to be estimated, and e is an n-vector of random errors incurred in observing y. The elements x_{ij} of X may have different statistical meanings depending on the application. We assume for simplicity that the error or noise e has expected value 0, and has covariance matrix $\sigma^2 V$, where V is a symmetric positive definite $n \times n$ matrix. Of course one can generalize to singular V and to unknown V and so on by using singular value decomposition and generalized inverses throughout to develop a more general theory, but I shall not do so here. I absorb the σ^2 or nonidentical row-dependent variances into V. A customary assumption on X is that $n \stackrel{>}{=} 2p$, i.e. one often thinks of X as having only a few (regressor) columns available. In fact it is useful to often think of p as just 1 or 2. Generally it seems to be usually assumed that the columns of X are linearly independent, and often it is assumed that those columns form an orthonormal set: $X'X = I_p$. For our considerations here I will assume the latter.

The Watson statistical efficiency for comparing an ordinary least squares estimator OLSE $\hat{\beta}$ and the best linear unbiased estimator BLUE β^* is defined as

$$\text{eff}\,(\hat{\beta}) = \frac{|\text{Cov}\,(\beta^*)|}{|\text{Cov}\,(\hat{\beta})|} = \frac{1}{|X'VX||X'V^{-1}X|}$$

where $|\cdot|$ denotes determinant. A fundamental lower bound for statistical efficiency is

$$\text{eff}\,(\hat{\beta}) \stackrel{>}{=} \prod_{i=1}^{p} \frac{4\lambda_i \lambda_{n-i+1}}{(\lambda_i + \lambda_{n-i+1})^2}$$

where $\lambda_1 \stackrel{>}{=} \lambda_2 \stackrel{>}{=} \cdots \stackrel{>}{=} \lambda_n > 0$ are the eigenvalues of V. This lower bound is sometimes called the Durbin–Bloomfield–Watson–Knott lower bound.

In [Gustafson (1999e)] the following new and geometrical interpretation of the lower bound was obtained. The essential meaning of this result is that the linear model's statistical efficiency is limited by the maximal turning angles of the covariance matrix V.

Theorem 6.1 (Trigonometry of Statistical Efficiency). *For the general linear model with SPD covariance matrix $V > 0$, for $p = 1$ the geometrical meaning of the relative efficiency of an OLSE estimator $\hat{\beta}$ against BLUE β^* is*

$$\text{eff}\,(\hat{\beta}) \stackrel{\geq}{=} \cos^2 \phi(V)$$

where $\phi(V)$ is the operator angle of V. For $p \stackrel{\leq}{=} n/2$ the geometrical meaning is

$$\text{eff}\,(\hat{\beta}) \stackrel{\geq}{=} \prod_{i=1}^{p} \cos^2 \phi_i(V) = \prod_{i=1}^{p} \mu_i^2(V)$$

where the $\phi_i(V)$ are the successive decreasing critical turning angles of V, i.e., corresponding to the higher antieigenvalues $\mu_i(V)$. The lower bound is attained for $p = 1$ by either of the two first antieigenvectors of V

$$x_\pm = \pm \left(\frac{\lambda_1}{\lambda_1 + \lambda_n} \right)^{1/2} x_n + \left(\frac{\lambda_n}{\lambda_1 + \lambda_n} \right)^{1/2} x_1.$$

For $p \stackrel{\leq}{=} n/2$ the lower bound is attained as

$$\prod_{i=1}^{p} \frac{\langle V x_\pm^i, x_\pm^i \rangle}{\|V x_\pm^i\| \|x_\pm^i\|}$$

where $x_\pm^i$ denotes the ith higher antieigenvectors of V given by

$$x_\pm^i = \pm \left(\frac{\lambda_i}{\lambda_i + \lambda_{n-i+1}} \right)^{1/2} x_{n-i+1} + \left(\frac{\lambda_{n-i+1}}{\lambda_i + \lambda_{n-i+1}} \right)^{1/2} x_i.$$

Bloomfield and Watson (1975) in their elegant treatment of statistical inefficiency also consider the commutators $C = XX'V - VXX'$ over all X such that $X'X = I_p$ where $2p \leq n$. A least squares estimation is optimal when $C = 0$ and as C becomes larger the estimation is poorer. To measure C the trace

$$\tau = \frac{1}{2} \text{tr}\,(C'C)$$

is used and they show that

$$\tau \stackrel{\leq}{=} \frac{1}{4} \sum_{i=1}^{p} (\lambda_i - \lambda_{n-i+1})^2,$$

where $p = \min\,(r, n - r)$, $r = \text{rank}\,(X)$.

Let us show how the operator trigonometry can quickly elaborate the meaning of such results. To obtain a "trigonometric" viewpoint, we write

$$\sum_{i=1}^{k} (\lambda_i - \lambda_{n-i+1})^2$$

$$= [(\lambda_1^2 + \lambda_n^2 - 2\lambda_1\lambda_n) + \cdots + (\lambda_k^2 + \lambda_{n-k+1}^2 - 2\lambda_k\lambda_{n-k+1})]$$

$$= \left[(\lambda_1^2 + \lambda_n^2) \left(1 - \frac{2\lambda_1\lambda_2}{\lambda_1^2 + \lambda_{n-k+1}^2} \right) + \cdots \right.$$

$$\left. + (\lambda_k^2 + \lambda_{n-k+1}^2) \left(1 - \frac{2\lambda_k\lambda_{n-k+1}}{\lambda_k^2 + \lambda_{n-k+1}^2} \right) \right]$$

$$= [(\lambda_1^2 + \lambda_n^2)(1 - \cos \phi_1(V^2)) + \cdots$$

$$+ (\lambda_k^2 + \lambda_{n-k+1}^2)(1 - \cos \phi_k(V^2))].$$

Here I have used (e.g., by the spectral mapping theorem) the fact that the cosine of the turning angle of the square of the covariance matrix V is $\cos \phi(V^2) = 2\lambda_1\lambda_n/(\lambda_1^2 + \lambda_n^2)$ and similarly for the other (i.e., smaller) critical turning angles of V^2. Thus the operator trigonometry shows the trace bound as one directly involving the partial trace of V^2 as reduced pairwise by its critical turning angles.

To get some explicit understanding of the Theorem 6.1 above, consider our matrix Examples 1, 2, and 3 in Chapter 1. If you happened to have a noise covariance matrix V in your statistics linear estimation problem that is scaled to one of three examples, then your efficiency eff $(\hat{\beta})$ would be bounded above by 1 and bounded below by the DBWK lower bounds, respectively,

$$\cos^2 \phi_1 = 0.888888889 \quad \text{(Example 1)},$$
$$\cos^2 \phi_1 = 0.9216 \quad \text{(Example 2)},$$

and for Example 3,

$$\cos^2 \phi_1 \cos^2 \phi_2 = \frac{4(20)}{(21)^2} \cdot \frac{4(20)}{(12)^2} = 0.100781053.$$

Notice that these bounds are uniform for whatever regressor (model) matrix X you may have implemented in your experiment.

To continue this example-led further explanation of the connection I established in the 1999 extension of my operator trigonometry to statistical efficiency, let me now refer to the quite recent paper [Chu, Isotalo, Puntanen, and Styan (2005)]. That paper goes into nice detail about efficiency and moreover presents some concrete examples. I have not discussed the following previously so you will need to look at their paper for more details. They consider the noise matrix

$$V = \begin{bmatrix} 3 & 1 & 1 & 3\rho \\ 1 & 3 & 1 & 1 \\ 1 & 1 & 3 & 1 \\ 3\rho & 1 & 1 & 3 \end{bmatrix}.$$

Thus $n = 4$ and $p = 2$. V is SPD whenever $1 > \rho > -2/3$. Let me take $\rho = 0$ here for simplicity and then work out on a preliminary basis some operator trigonometry and then return to their considerations of Watson factorized efficiencies.

First we need to calculate the eigenvalues of V with ρ taken equal to zero. Its characteristic equation becomes the calculation

$$0 = \begin{vmatrix} (3-\lambda) & 1 & 1 & 0 \\ 1 & (3-\lambda) & 1 & 1 \\ 1 & 1 & (3-\lambda) & 1 \\ 0 & 1 & 1 & (3-\lambda) \end{vmatrix}$$

$$= (3-\lambda)\begin{vmatrix} (3-\lambda) & 1 & 1 \\ 1 & (3-\lambda) & 1 \\ 1 & 1 & (3-\lambda) \end{vmatrix} - \begin{vmatrix} 1 & 1 & 1 \\ 1 & (3-\lambda) & 1 \\ 0 & 1 & (3-\lambda) \end{vmatrix}$$

$$+ \begin{vmatrix} 1 & (3-\lambda) & 1 \\ 1 & 1 & 1 \\ 0 & 1 & (3-\lambda) \end{vmatrix}$$

$$= (3-\lambda)[(3-\lambda)^2 + 2 - 3(3-\lambda)] - [(3-\lambda)^2 + 1 - ((3-\lambda)+1)]$$

$$+ [(3-\lambda)+1 - ((3-\lambda)^2 + 1)]$$

$$= (3-\lambda)^4 - 5(3-\lambda)^2 + 4(3-\lambda).$$

Thus one eigenvalue is $\lambda = 3$. Replacing $3 - \lambda$ with x we need to factor the cubic equation $x^3 - 5x + 4 = 0$ and see that $x = 1$, and hence $\lambda = 2$ will be one root. That leaves the quadratic equation $x^2 + x - 4 = 0$ from which we find the other two eigenvalues of V from the roots $x = (-1 \pm \sqrt{17})/2$. Thus

V has the four eigenvalues

$$\lambda_1 = 5.561552813,$$
$$\lambda_2 = 3,$$
$$\lambda_3 = 2,$$
$$\lambda_4 = 1.438447187.$$

From these we could now find the critical turning angles ϕ_1 and ϕ_2 from their cosines as we did in the 4×4 Example 3 considered above. But alternately we may use

$$\sin \phi_1(\mathbf{A}) = \frac{\lambda_1 - \lambda_4}{\lambda_1 + \lambda_4}$$

which is simpler arithmetically. For the noise matrix **V** we have

$$\frac{\lambda_1 - \lambda_4}{\lambda_1 + \lambda_4} = \frac{4.123105626}{7} = 0.589015089 = \sin \phi_1(\mathbf{V})$$

and hence the maximal turning angle $\phi_1(\mathbf{V})$ is

$$\phi_{14} = 36.08714707 \text{ degrees.}$$

From this we immediately obtain

$$\cos \phi_{14} = 0.808122036,$$
$$\cos^2 \phi_{14} = 6.653061224.$$

In the same manner, from

$$\frac{\lambda_2 - \lambda_3}{\lambda_2 + \lambda_3} = \frac{1}{5} = \sin \phi_2(\mathbf{V})$$

we have the second (interior) critical turning angle $\phi_2(\mathbf{V})$

$$\phi_{23} = 11.53695903 \text{ degrees}$$

and

$$\cos \phi_{23} = 0.979795897,$$
$$\cos^2 \phi_{23} = 0.96.$$

Hence we may compute the DBWK lower bound

$$\text{eff} (\hat{\beta}) \overset{\geq}{=} \cos^2 \phi_1(\mathbf{V}) \cos^2 \phi_2(\mathbf{V})$$

$$= 0.626938775.$$

Notice that so far the regressor matrix **X** has not yet entered the picture.

Chu, Isotalo, Puntanen, and Styan (2005) now bring in two examples with specific partitioned $\mathbf{X} = (\mathbf{X}_1 : \mathbf{X}_2)$ and factorize the efficiency. So let us continue illustrating our operator trigonometric connection to statistical efficiency by using their specific examples for $\mathbf{X}$ for the Model Problem

$$y = \mathbf{X}_1 \beta_1 + \mathbf{X}_2 \beta_2 + \epsilon$$

where noise covariance matrix is the $\mathbf{V}$ above. Without going into all their analysis of efficiency factorization multipliers, etc., we may go directly to their first example

$$\mathbf{V} = [\mathbf{X}_1 : \mathbf{X}_2] = \begin{bmatrix} 1 : -1 \\ 2 : -2 \\ 1 : +2 \\ 1 : +1 \end{bmatrix}.$$

They find a factorized total efficiency

$$\text{eff}\,(\hat{\beta}) = \text{eff}\,(\hat{\beta}_1)\,\text{eff}\,(\hat{\beta}_2)$$

$$= \frac{400(2 + 3\rho)(1 - \rho)}{(1 + \rho)(11 + 3\rho)(7 - 6\rho)(11 - 3\rho)}$$

which in our case with $\rho = 0$ becomes

$$\text{eff}\,(\hat{\beta}, \text{given } \mathbf{X}) = \frac{800}{7(121)} = 0.944510035,$$

indicating an estimation accuracy quite a bit better than our worst case DBWK lower bound 0.626938775 obtained above. Their second example is

$$\mathbf{X} = [\mathbf{X}_1 : \mathbf{X}_2] = \begin{bmatrix} 1 : -1 \\ 1 : +1 \\ 1 : +2 \\ 1 : -2 \end{bmatrix}$$

and they arrive at a total efficiency expression

$$\text{eff}\,(\hat{\beta}, \text{given } \mathbf{X}) = \frac{12800(3\rho + 2)(1 - \rho)}{(77 - 70\rho - 3\rho^2)(343 + 414\rho - 9\rho^2)}$$

$$= \frac{25600}{(77)(343)} = 0.969293098$$

when $\rho = 0$, a bit better than with the first regressor matrix $\mathbf{X}$.

From the DBWK inefficiency bound, it is clear that your worse choice regressors would occur if you happened to choose data which gave you exactly all of the antieigenvector pairs $x_{\pm}^k$.

There is much more inefficiency theory if one considers also noninvertible noise matrices $\mathbf{V}$. One treats those by the theories of generalized inverses $\mathbf{V}^-$, e.g., as in Moore–Penrose pseudoinverse theory. The matrix statistics community is quite advanced in this direction. But I am not, and stay away from it when I can, though the $\mathbf{V}^-$ theory is very useful.

On the other hand, for some reason I felt compelled to know more about the history of this general linear model which is so basic to matrix statistics. In [Gustafson (1999e; 2002a)] I followed it back to the contributions of Gauss and Markov, among others. Then in [Gustafson (2005a)] I followed the consequent inefficiency theory back to Fisher and Student. But my most interesting find was a connection to Von Neumann. It is possible that nobody alive today would happen to know that connection which so interests me. So I want to give it here.

Tracing back through the two papers [Durbin and Watson (1950; 1951)], one comes upon the interesting $n \times n$ matrix

$$
A = \begin{bmatrix}
1 & -1 & 0 & \cdots & & \cdot & 0 \\
-1 & 2 & -1 & 0 & & \cdot & 0 \\
0 & -1 & 2 & -1 & & \cdot & 0 \\
\vdots & & \cdot & \cdot & \cdot & & 0 \\
& & & \cdot & -1 & 2 & -1 \\
0 & \cdots & & & 0 & -1 & 1
\end{bmatrix}.
$$

It is stated there that this results from the statistic for testing serial correlation

$$
d = \frac{\sum (\Delta z)^2}{\sum z^2} = \frac{\langle Az, z \rangle}{\sum z^2},
$$

where z is the residual from linear regression. It was shown (1951) that the mean and variance of the statistic d are given by

$$
E(d) = \frac{P}{n - k' - 1},
$$

$$
\text{var}(d) = \frac{2[Q - PE(d)]}{(n - k' - 1)(n - k' + 1)},
$$

where

$$P = \text{tr}\, A - \text{tr}(X'AX(X'X)^{-1}),$$

$$Q = \text{tr}\, A^2 - 2\,\text{tr}(X'A^2X(X'X)^{-1}) + \text{tr}(X'AX(X'X)^{-1})^2,$$

where k' is the number of columns of the matrix of observations of the independent variables

$$\begin{bmatrix} x_{11} & x_{21} & \cdots & x_{k'1} \\ \vdots & & & \\ x_{1n} & x_{2n} & \cdots & x_{k'n} \end{bmatrix}.$$

One wondered, or at least this author wondered, how A came about.

A more careful reading of Durbin and Watson (1950) leads to a paper of J. Von Neumann (1941) and one cannot resist looking at it. As is well-known, Von Neumann was a polymath and this paper is no exception. An in-depth study of the statistic

$$\eta = \frac{\delta^2}{s^2}$$

is carried out, where s^2 is the sample variance of a normally distributed random variable and $\delta^2 = \sum_{\mu=1}^{n-1} (x_{\mu+1} - x_\mu)^2/(n-1)$ is the mean square successive difference, the goal being to determine the independence or trend dependence of the observations $x_1, \ldots, x_n$. Thus we find this paper to be an early and key precedent to all the work by Durbin, Watson, and others in the period 1950–1975.

Von Neumann's analysis is extensive and he obtains a number of theoretical results which, if we might paraphrase Durbin and Watson (1950), p. 418, are more or less beyond use by conventional statisticians. However, both Durbin–Watson papers [Durbin and Watson (1950; 1951)] go ahead and use the matrix A to illustrate their theory. So one looks further into Von Neumann's paper to better understand the origin of the matrix A. One finds there (p. 367) the statement: "The reasons for the study of the distribution of the mean square successive difference δ^2, in itself as well as in its relationship to the variance s^2, have been set forth in a previous publication, to which the reader is referred." However it is made clear that comparing observed values of the statistic η will be used to determine "whether the observations $x_1, \ldots, x_n$ are independent or whether a trend exists."

Curiosity knowing no bounds, I pushed the historical trace back to the previous publication [Von Neumann, Kent, Bellison, and Hart (1941)]. The answer to our curiosity about why Von Neumann became involved with this statistical regression problem is found there. To quote (p. 154): "The usefulness of the differences between successive observations only appears to have been realized first by ballisticians, who faced the problem of minimizing effects due to wind variation, heat and wear in measuring the dispersion of the distance traveled by shell." That paper originated from the Aberdeen Ballistic Research Laboratory, where Von Neumann was consulting.

Returning to his analysis in Von Neumann (1941), we find he begins with a now more or less classical multivariate analysis of normally distributed variables. By diagonalization, a quadratic form $\sum A_\mu x'_\mu$ is obtained where the A_μ, $\mu = 1, \ldots, n$, are the eigenvalues of the form $(n-1)\delta^2$. A smallest eigenvalue $A_n = 0$ is found, with eigenvector $x_0 = (1, \ldots, 1)/\sqrt{n}$. A further analysis, using an interesting technique by assuming the $x'_1, \ldots, x'_{n-1}$ to be uniformly distributed over an $(n-1)$-unit sphere, shows that the statistic η is then distributed according to

$$\eta = \frac{n}{n-1} \sum_{\mu=1}^{n-1} A_\mu x_\mu^2.$$

Thus the sought eigenvalues A_μ, $\mu = 1, \ldots, n$, are the eigenvalues of the quadratic form $(n-1)\delta^2$, which is then written as

$$(n-1)\delta^2 = x_1^2 + 2\sum_{\mu=2}^{n-1} x_\mu^2 + x_n^2 - 2\sum_{\mu=1}^{n-1} x_\mu x_{\mu+1}.$$

The matrix of this form is A and it is that matrix which is also borrowed and used in [Durbin and Watson (1950; 1951)]. Used as well are the eigenvalues

$$A_k = 4\sin^2\left(\frac{k\pi}{2n}\right), \quad k = 0, 1, \ldots, n-1$$

which Von Neumann computes from the determinant of A.

When I first saw the matrix A in [Durbin and Watson (1950; 1951)], my take was completely different. As this author is a specialist in partial differential equations, I immediately saw the matrix A as the discretized

Poisson–Neumann boundary value problem

$$\begin{cases} -\dfrac{d^2u(x)}{dx^2} = f(x), & 0 < x < 1, \\ \dfrac{du}{dx} = 0 & \text{at } x = 0, 1. \end{cases}$$

In saying this I am disregarding the exact interval and discrete Δx sizes.

This new connection between statistical efficiency and partial differential equations should be further explored elsewhere, especially as it will no doubt generalize to Dirichlet, Neumann, and Robin boundary value problems for the Laplacian operator $-\Delta = \sum \partial^2 u/\partial x^2$ in higher dimensions. The reverse implications to induce a more general context of statistical efficiency would seem to be interesting.

We also comment in passing that a similar ballistics problem, that of control of rocket flight, was the motivating application in Japan during the Second World War that led Ito to develop his stochastic calculus now so important in the theory of financial derivatives and elsewhere. See Exercise 1.

6.2   The Euler Equation versus the Inefficiency Equation

The fundamental DBWK lower bound on relative least squares efficiencies is usually proved in the statistics literature by means of highly developed Lagrangian variational methods. Following the book [Wang and Chow (1994)], we see that it is easier to just work with the denominator of the efficiency eff $(\hat{\beta})$, that is,

$$(\text{eff}\,(\hat{\beta}))^{-1} = |XV^{-1}X\|X'VX|,$$

the general case having been reduced to that of $X'X = I_p$. By a differentiation of $F(x, \lambda) = \ln|X'V^{-1}X| + \ln|X'VX| - 2\,\text{tr}(X'X\Lambda)$ and subsequent minimization, the relation

$$X'X(\Lambda + \Lambda') = \Lambda + \Lambda' = 2I_p$$

is obtained. Here Λ is a $p \times p$ upper triangular matrix which is the Lagrange multiplier with respect to the constant $X'X = I_p$. From this and further work including the simultaneous diagonalization of $X'V^2X$, $X'VX$, and $X'V^{-1}X$, one arrives at the result

$$(\text{eff}(\hat{\beta})^{-1}) = \prod_{t=1}^{p} x_i' V x_i x_i' V^{-1} x_i$$

where X is now the $n \times p$ column matrix $X = [(x_1) \cdots (x_p)]$ whose columns go into this expression. The Lagrange multiplier minimization has also now yielded the equation for the x_i:

$$\frac{V^2 x_i}{x_i' V x_i} + \frac{x_i}{x_i' V^{-1} x_i} = 2 V x_i, \quad i = 1, \ldots, p.$$

Clearly the span $\{x_i, V x_i\}$ is a two- (or one-)dimensional reducing subspace of V and is spanned by two (or one) eigenvectors ψ_j and ψ_k of V. Writing each column $x_i = \sum_{j=1}^{n} \alpha_{ij} \psi_j$ in terms of the full eigenvector basis of V yields the quadratic equation

$$\frac{z^2}{x_i' V x_i} - 2z + \frac{1}{x_k' V^{-1} x_i} = 0$$

for the two (or one) eigenvalues λ_j and λ_k associated to each $x_i, i = 1, \ldots, p$. Substituting those eigenvalues as found from this last equation into the p-product expression for $(\text{eff}(\hat{\beta}))^{-1}$ brings that expression to the DBWK lower bound.

I called the above equation for the columns x_i of the regression design matrices X the inefficiency equation. Let me rewrite it here in our operator theory notation,

$$\frac{V^2 x}{\langle V x, x \rangle} - 2 V x + \frac{x}{\langle V^{-1} x, x \rangle} = 0.$$

This equation clearly invites comparison to our operator trigonometry Euler equation

$$\frac{A^2 x}{\langle A^2 x, x \rangle} - \frac{2 A x}{\langle A x, x \rangle} + x = 0.$$

In both of these equations I have assumed the operators to be SPD and the vectors of norm one.

Theorem 6.2 (Inefficiency Equation versus Euler Equation). *For any $n \times n$ SPD covariance matrix V or more generally any $n \times n$ SPD matrix A, all* *eigenvectors x_j satisfy the inefficiency equation and the Euler equation. The only other vectors satisfying the inefficiency equation are the "inefficiency vectors"*

$$x_{\pm}^{j,k} = \pm \frac{1}{\sqrt{2}} x_j + \frac{1}{\sqrt{2}} x_k,$$

where x_j and x_k are any eigenvectors corresponding to any distinct eigenvalues $\lambda_j \neq \lambda_k$. The only other vectors satisfying the Euler equation are the antieigenvectors

$$x_{\pm}^{jk} = \pm \left(\frac{\lambda_k}{\lambda_j + \lambda_k} \right)^{1/2} x_j + \left(\frac{\lambda_j}{\lambda_j + \lambda_k} \right)^{1/2} x_k.$$

The proof of this theorem was given in [Gustafson (1999e; 2002a)]. It goes as follows. For simplicity we regard all eigenvalues λ_j to be distinct.

First, that V's and A's eigenvectors x_j all satisfy both equations follows immediately by direct substitution. For the next step we prefer to write the equations as

$$V^2 x - 2\langle Vx, x \rangle Vx + \langle Vx, x \rangle \langle V^{-1}x, x \rangle^{-1} x = 0$$

and

$$A^2 x - 2\langle A^2 x, x \rangle \langle Ax, x \rangle^{-1} Ax + \langle A^2 x, x \rangle x = 0.$$

Let x, $\|x\| = 1$, be any noneigenvector solution of the inefficiency equation. By the reducing subspace argument (see Exercise 3) we know that

$$x = c_j x_j + c_k x_k$$

for some eigenvectors x_j and x_k corresponding to some eigenvalues $\lambda_j \neq \lambda_k$. We note that $Vx = c_j \lambda_j x_j + c_k \lambda_k x_k$, $\langle Vx, x \rangle = c_j^2 \lambda_j + c_k^2 \lambda_k$, similarly $\langle V^{-1}x, x \rangle = c_j^2 \lambda_j^{-1} + c_k^2 \lambda_k^{-1}$. The eigenvalues must satisfy the quadratic equation

$$z^2 - 2\langle Vx, x \rangle z + \langle Vx, x \rangle \langle V^{-1}x, x \rangle^{-1} = 0.$$

Because the two roots λ_1 and λ_2 to any polynomial equation $\lambda^2 - a_1 \lambda + a_2 = 0$ always satisfy $\lambda_1 + \lambda_2 = a_1$ and $\lambda_1 \lambda_2 = a_2$, we therefore have of necessity that

$$\langle Vx, x \rangle = \frac{(\lambda_j + \lambda_k)}{2}, \quad \langle V^{-1}x, x \rangle = \frac{(\lambda_j + \lambda_k)}{2\lambda_j \lambda_k}.$$

Thus the coefficients c_j and c_k must satisfy the system

$$c_j^2 \lambda_j + c_k^2 \lambda_k = \frac{\lambda_j + \lambda_k}{2}$$

$$c_j^2 \lambda_j^{-1} + c_k^2 \lambda_k^{-1} = \frac{\lambda_j + \lambda_k}{2\lambda_j \lambda_k}.$$

Solving this system by Cramer's rule yields

$$c_j^2 = \begin{vmatrix} (\lambda_j + \lambda_k)/2 & \lambda_k \\ (\lambda_j + \lambda_k)/2\lambda_j\lambda_k & \lambda_k^{-1} \end{vmatrix} \Bigg/ \begin{vmatrix} \lambda_j & \lambda_k \\ \lambda_j^{-1} & \lambda_k^{-1} \end{vmatrix} = \frac{\left(\frac{\lambda_j+\lambda_k}{2}\right)(\lambda_k^{-1} - \lambda_j^{-1})}{(\lambda_j^2 - \lambda_k^2)/\lambda_j\lambda_k} = \frac{1}{2}.$$

Similarly $c_k^2 = 1/2$ and thus the only noneigenvector solutions to the inefficiency equation are those given in the theorem.

We treat the Euler equation in the same way. Again the reducing subspace argument allows us to write any noneigenvector solution $x = c_j x_j + c_k x_k$ and we arrive at the quadratic equation

$$z^2 - 2\frac{\langle A^2 x, x\rangle}{\langle Ax, x\rangle}z + \langle A^2 x, x\rangle = 0$$

for two eigenvalues $\lambda_j \neq \lambda_k$. Necessarily

$$2\langle A^2 x, x\rangle \langle Ax, x\rangle^{-1} = \lambda_j + \lambda_k, \quad \langle A^2 x, x\rangle = \lambda_j \lambda_k.$$

The coefficients c_j and c_k therefore must satisfy the system

$$\begin{bmatrix} \lambda_j & -\lambda_k \\ \lambda_j^2 & \lambda_k^2 \end{bmatrix} \begin{bmatrix} c_j^2 \\ c_k^2 \end{bmatrix} = \begin{bmatrix} 0 \\ \lambda_j \lambda_k \end{bmatrix}.$$

By Cramer's rule we have

$$c_j^2 = \frac{\begin{vmatrix} 0 & -\lambda_k \\ \lambda_j\lambda_k & \lambda_k^2 \end{vmatrix}}{(\lambda_j\lambda_k)(\lambda_j - \lambda_k)} = \frac{\lambda_j\lambda_k^2}{\lambda_j\lambda_k(\lambda_j + \lambda_k)} = \frac{\lambda_k}{\lambda_j + \lambda_k}.$$

Similarly $c_k^2 = \lambda_j/(\lambda_j + \lambda_k)$ and thus the only noneigenvector solutions to the Euler equation are those given in the theorem.

A few things may be noted here.

First, this argument provides another way to see the two-component nature of the antieigenvectors which I had earlier obtained from proving my Min-Max Theorem [Gustafson (1968d; 1972a)], see Sec. 3.2. It interests me that Bloomfield and Watson in (1975) used a Lagrange multiplier argument to obtain a two-component solution to the inefficiency equation, which is, after all, an Euler equation for a different but related variational functional. Second, one need not use either my Min-Max quadratic convex minimization arguments nor the Lagrange multiplier arguments to solve for the specific coefficients: they can be found by direct computation as I did in Sec. 3.4. Third, I hinted at the end of Sec. 3.4 that there could be further interesting "Morse theory" to be explored for these kinds of variational functionals, and for that one might wish to allow three and higher component achieving vectors. A mountain range has more than tops and bottoms.

6.3   Canonical Correlations and Rayleigh Quotients

In this section we consider canonical correlations within a context of Rayleigh quotients. The term canonical correlations is used variously within the statistics literature.

The first situation is that of C. R. Rao and C. V. Rao (1987) who are interested in what they call homologous (i.e., implicitly related) measurements from family members. They are led to consider stationary values of an expression

$$\frac{x'Cx}{(x'Ax)^{1/2}(x'Bx)^{1/2}}$$

with A and B symmetric positive definite and C symmetric. Squaring this gives the product of two Rayleigh quotients

$$\frac{\langle Cx, x \rangle}{\langle Ax, x \rangle} \cdot \frac{\langle Cx, x \rangle}{\langle Bx, x \rangle}.$$

Taking the functional derivative of the original expression with respect to x yields the equation

$$\frac{x'Cx}{x'Ax}Ax + \frac{x'Cx}{x'Bx}Bx = 2Cx.$$

Note that if we let $C = T$, $A = T^2$, $B = I$, then the first expression becomes our antieigenvalue quotient. Similarly the third expression for the same operators x normalized to $\|x\| = 1$ becomes the Euler equation. Thus a general theory encompassing statistical efficiency, operator trigonometry, and such canonical correlations, could be developed.

In their analysis, C. R. Rao and C. V. Rao (1987) arrive at two cases, the first corresponding to stationary values equal to 1, the second corresponding to smaller stationary values. As concerns the second case, they note that "there can be solutions of the form $x = ae_i + be_j$," where the e_i and e_j are eigenvectors. But we now know from the operator trigonometry that these are the two cases covered by our Euler equation, and that the solutions in the second case are analogous to our antieigenvectors.

Interestingly, the Raos were led to consider such variational quotients from earlier work in electrical engineering by [Cameron and Kouvaritakis (1980)], see also [Cameron (1983)]. So if we look at the latter more mathematically-stated paper, we find the general variational quotient

$$\mu(x) = \frac{\langle Ax, x\rangle \langle Bx, x\rangle}{\langle x, x\rangle \langle x, x\rangle}$$

leading to two equations for its stationary points, namely

$$\frac{1}{2}\left\{\frac{\langle Bx, x\rangle}{\langle x, x\rangle}A + \frac{\langle Ax, x\rangle}{\langle x, x\rangle}B\right\}x = \mu(x)$$

or

$$\frac{1}{2}\left\{\frac{A}{\langle Ax, x\rangle} + \frac{B}{\langle Bx, x\rangle}\right\}x = \frac{x}{\langle x, x\rangle}.$$

From what we have seen in the previous section, intuition would say that the first equation above corresponds to μ being analogous to an eigenvalue, and the second equation above corresponds to x being analogous to an antieigenvector.

I have not explored such theory nor its potential application to control theory or antenna design. An interesting observation in the second paper is that any solution of the second equation must bisect the angle between Ax and Bx. If so, then x would resemble the equal weighting one has in the inefficiency vectors. I do not have time to go further into this application area, but do suggest in Exercise 4 that it might be interesting to do so. I have given a few hints there.

Therefore a rather general perspective to conclude this section would be to conjecture a future general theory which encompasses all three of the variational theories of the Euler equation of the matrix trigonometry, the Lagrangian and other functional differentiation methods of the statistical efficiency theory, and as many relevant results and methods as appropriate from control theory. The matrix trigonometry will give the geometry, the Lagrangian methods the analysis, and the control theory the physical or statistical applications.

6.4 Other Statistics Inequalities

The recent outstanding survey [Drury et al. (2002)] on matrix inequalities in statistics allows me to bring out some more matrix statistics trigonometry here. These were reported in [Gustafson (2006b)]. The reader will probably want to refer to the [Drury et al. (2002)] survey for more details about the following. In particular, you will find the bibliographical citations there for some papers I refer to here.

The [Khatri–Rao (1981; 1982)] inequality generalizes the Bloomfield–Watson–Knott efficiency inequality to singular values rather than eigenvalues. One version is [Drury et al. (2002), p. 494, (6.2)]

$$\text{tr}\,(T'KU \cdot U'K^{-1}T) \stackrel{\le}{=} \sum_{i=1}^{m} \frac{(\sigma_i + \sigma_{n-i+1})^2}{4\sigma_i\sigma_{n-i+1}} + p - m.$$

Here T and U are $n \times p$ matrices, $T'T = U'U = I_p$, K is an $n \times n$ nonsingular matrix with singular values $\sigma_1 \ge \sigma_2 \ge \cdots \ge \sigma_n > 0$, and $m = \min(p, n - p)$. Of course the singular values are just the eigenvalues of the SPD polar factor $|K| = (K'K)^{1/2}$, so my extended operator trigonometry [Gustafson (2000d)] identifies the sum on the right-hand side as $\sum_{i=1}^{m} \cos^{-2} \phi_i(|K|)$. In other words, the bound is trigonometric.

Going further, Ando (2001) proved [Drury et al. (2002), p. 494, (6.7)]

$$\{\text{ch}_i(A_{21}A_{11}^{-2}A_{21})\}_{i=1}^{m} \prec_w \left\{ \frac{(\lambda_i - \lambda_{n-i+1})^2}{4\lambda_i\lambda_{n-i+1}} \right\}_{i=1}^{m}.$$

Here the "weakly majorized" symbol $\{\alpha_i\} \prec_w \{\beta_i\}$ means that $\sum_1^h \alpha_i \stackrel{\le}{=} \sum_1^h \beta_i$ for all $h = 1, \ldots, m$, and the ch_i means the ith largest eigenvalue. Due to the minus signs the right-hand side is not yet trigonometric, but we may make it so. See [Gustafson (2006b)] for details, from which this Khatri–Rao–Ando bound is trigonometric:

$$\{\text{ch}_i(A_{21}A_{11}^{-2}A_{21})\}_{i=1}^{m} \prec_w \{\tan^2 \phi_i(A)\}_{i=1}^{m}.$$

Here the $\phi_i(A)$ are the matrix trigonometric nested turning angles of A.

The generalized [Bartmann–Bloomfield (1981)] determinant lower bound [see Drury et al. (2002), (5.8) on p. 480]

$$|I - (X'AX)^- X'AY(Y'AY)^- Y'AX| = \prod_{i=1}^{h} (1 - \kappa_i^2) \stackrel{\ge}{=} \prod_{i=1}^{\min(r_1, r_2)} \frac{4\lambda_i\lambda_{r-i+1}}{(\lambda_i + \lambda_{r-i+1})^2}$$

is also clearly trigonometric. Here A is a dispersion matrix with rank $r \leqq n$ and eigenvalues $\lambda_1 \geqq \lambda_2 \geqq \cdots \geqq \lambda_r$, and κ_i denotes the canonical correlations. The exact definitions of the canonical correlations, depending on the generality desired, may be found in [Drury *et al.* (2002)]. Another way to describe canonical correlations is found in the important early paper [Eaton (1976)], which Bartmann and Bloomfield (1981) extend. There, the largest canonical correlation is called $\theta_1 = \sup \langle A_{12}x, y \rangle / \langle A_{11}y, y \rangle^{1/2} \langle A_{22}x, x \rangle^{1/2}$ over all $x, y \neq 0$. All canonical correlations can thus be defined in terms of matrix partitions. Eaton showed that $\theta_1 \leqq (\lambda_1 - \lambda_n)/(\lambda_1 + \lambda_n)$, which we may now know as: $\theta_1 \leqq \sin \phi(A)$. More generally, without giving all details here, the $(1 - \kappa_i^2)$ factors will mean, trigonometrically now: $\kappa_i^2 \leqq \sin^2 \phi_i(A)$. Thus $1 - \kappa_i^2 \geqq 1 - \sin^2 \phi_i(A) = \cos^2 \phi_i(A)$, hence giving the lower bound above.

Not always does a geometrical meaning via our matrix trigonometry come out so cleanly. An example is the conjecture (7.6) in [Drury *et al.* (2002)], that the canonical correlations satisfy

$$\prod_{i=1}^{k}(1 - \kappa_i) \geqq \prod_{i=1}^{k}\left(1 - \frac{\lambda_i - \lambda_{n-i+1}}{\lambda_i + \lambda_{n-i+1}}\right),$$

$k = 1, \ldots, m$. The right-hand side is nicely trigonometric, the factors being $1 - \sin \phi_i(A)$ where the ϕ_i are the turning angles of A. But what does it mean geometrically? It would be nice if an answer to that could provide a proof or counterexample to the conjecture (7.6) in [Drury *et al.* (2002)].

More hopeful is a trigonometric proof of the bound [Drury *et al.* (2002), (5.40) on p. 486] for the Hotelling correlation coefficient $\rho_2 = \prod_{i=1}^{m} \kappa_i^2$ for an SPD $n \times n$ dispersion matrix A, namely

$$\prod_{i=1}^{m}\kappa_i^2 \leqq \prod_{i=1}^{m}\left(\frac{\lambda_i - \lambda_{n-i+1}}{\lambda_i + \lambda_{n-i+1}}\right)^2 \leqq \prod_{i=1}^{m}\left(\frac{\lambda_i - \lambda_{n-m+i}}{\lambda_i + \lambda_{n-m+i}}\right)^2 \leqq \left(\frac{\lambda_1 - \lambda_n}{\lambda_1 + \lambda_n}\right)^n.$$

The far right inequality is obvious and furthermore we note that the bound is exactly in $\sin^n \phi(A)$. However, the authors (p. 486) state "we find the left-hand inequalities to be surprising!"

The proof in [Drury *et al.* (2002)] is quite involved and depends on an inspection of a whole set of permutations. Let us consider, as they do, the special case $n = 4$, $m = 2 = n - p$, so that becomes [Drury *et al.* (2002), by (5.41)] the inequalities

$$\left(\frac{\lambda_1 - \lambda_4}{\lambda_1 + \lambda_4}\right)^2 \left(\frac{\lambda_2 - \lambda_3}{\lambda_2 + \lambda_3}\right)^2 \leq \left(\frac{\lambda_1 - \lambda_3}{\lambda_1 + \lambda_3}\right)^2 \left(\frac{\lambda_2 - \lambda_4}{\lambda_2 + \lambda_4}\right)^2 \leq \left(\frac{\lambda_1 - \lambda_4}{\lambda_1 + \lambda_4}\right)^4.$$

With $\lambda_1 > \lambda_2 > \lambda_3 > \lambda_4 > 0$ for simplicity, we recognize the left-hand part as the matrix trigonometric inequality

$$\sin \phi_{14} \sin \phi_{23} \overset{\leq}{=} \sin \phi_{13} \sin \phi_{24}$$

where $\phi_{14} = \phi(A)$, $\phi_{23} = \phi_2(A)$ are the critical turning angles of A. The ϕ_{13} and ϕ_{24} would be the turning angles of A determined by eigenvectors x_1 and x_3, x_2 and x_4, respectively. From this observation, I wondered if one could give a purely trigonometric proof of the Hotelling correlation coefficient bound. Indeed, that can be done.

Theorem 6.3. (The Hotelling Correlation Bound is Trigonometric). *Moreover, it may be derived operator trigonometrically.*

Proof. First let us prove the special case. That inequality is easily seen to be equivalent to the inequality

$$\frac{1 - \left(\frac{\lambda_1\lambda_3 + \lambda_2\lambda_4}{\lambda_1\lambda_2 + \lambda_3\lambda_4}\right)}{1 + \left(\frac{\lambda_1\lambda_2 + \lambda_3\lambda_4}{\lambda_1\lambda_2 + \lambda_3\lambda_4}\right)} \overset{\leq}{=} \frac{1 - \left(\frac{\lambda_1\lambda_4 + \lambda_2\lambda_3}{\lambda_1\lambda_2 + \lambda_3\lambda_4}\right)}{1 + \left(\frac{\lambda_1\lambda_4 + \lambda_2\lambda_3}{\lambda_1\lambda_2 + \lambda_3\lambda_4}\right)}.$$

This expression is of the form $f(t_1) \overset{\leq}{=} f(t_2)$ where $f(t) = (1 - t)/(1 + t)$ is strictly decreasing for $t > -1$. Because $\lambda_1\lambda_3 + \lambda_2\lambda_4 > \lambda_1\lambda_4 + \lambda_2\lambda_3$ which is because $\lambda_3 > \lambda_4$, we have a proof by working backward from these relations. But here we want this to be seen in an operator trigonometric geometric way. Not immediately seeing how to show and proceed directly, instead we see that the above expression with its common denominator can be rewritten as

$$\sin \phi(A_{\text{LHS}}) \overset{\leq}{=} \sin \phi(A_{\text{RHS}})$$

where A_{LHS} is an imagined operator with largest eigenvalue $\lambda_1\lambda_2 + \lambda_3\lambda_4$ and smallest eigenvalue $\lambda_1\lambda_3 + \lambda_2\lambda_4$, and similarly A_{RHS} has largest eigenvalue $\lambda_1\lambda_2 + \lambda_3\lambda_4$ and smallest eigenvalue $\lambda_1\lambda_4 + \lambda_2\lambda_3$. Working backward from this sine relation now proves the Hotelling bound in the special case. The geometrical meaning of the Hotelling bound thus is that A_{LHS} has smaller turning angle than A_{RHS}.

Now let us consider the general case. But that reduces to the special case. To see that, consider the case $n = 8$, $m = 4$. Then we want

$$\left(\frac{\lambda_1 - \lambda_8}{\lambda_1 + \lambda_8}\right)^2 \left(\frac{\lambda_2 - \lambda_7}{\lambda_2 + \lambda_7}\right)^2 \left(\frac{\lambda_3 - \lambda_6}{\lambda_3 + \lambda_6}\right)^2 \left(\frac{\lambda_4 - \lambda_5}{\lambda_4 + \lambda_5}\right)^2$$

$$\stackrel{\leq}{=} \left(\frac{\lambda_1 - \lambda_5}{\lambda_1 + \lambda_5}\right)^2 \left(\frac{\lambda_2 - \lambda_6}{\lambda_2 + \lambda_6}\right)^2 \left(\frac{\lambda_3 - \lambda_7}{\lambda_3 + \lambda_7}\right)^2 \left(\frac{\lambda_1 - \lambda_8}{\lambda_1 + \lambda_8}\right)^2.$$

Let us first select the two "outside" factors of both the left-hand side and right-hand side of this inequality. They then satisfy our previously shown special case. Then we select the two "inside" factors of both the left-hand side and right-hand side. They also satisfy the spectral ordering needed for the special case. Hence for $n = 2m$, one may proceed from the outside in, or the inside out. In the case that m is odd, there is a factor repeated on both sides.

I especially like this operator trigonometric proof of a hard inequality in statistics. To illustrate the geometric meaning of that inequality and my new trigonometry for it, I have sketched in Fig. 6.1(a) the four $\sin_{ij}(A)$ convex minima for the special case

$$A = \begin{bmatrix} \lambda_4 & & & \\ & \lambda_3 & & \\ & & \lambda_2 & \\ & & & \lambda_1 \end{bmatrix} = \begin{bmatrix} 1 & & & \\ & 2 & & \\ & & 3 & \\ & & & 4 \end{bmatrix} \sim \begin{bmatrix} 1/4 & & & \\ & 1/3 & & \\ & & 1/2 & \\ & & & 1 \end{bmatrix}.$$

That is, since the operator trigonometry of A does not change under rescalings, the above fractional-eigenvalue version provides a better visualization. Then, as in Fig. 1.1, line $\ell_1(+1)$ is $1 - \epsilon\lambda_1$ and line $\ell_4(-1)$ is $-1 + \epsilon\lambda_4$ and those form the $\|\epsilon A - I\|$ curve. Similarly the other intersecting lines create all of the desired convex minima. Thus for the inequality $\sin\phi_{14}\sin\phi_{23} \stackrel{\leq}{=} \sin\phi_{13}\sin\phi_{24}$, expressed as

$$\frac{\lambda_1 - \lambda_4}{\lambda_1 + \lambda_4} \cdot \frac{\lambda_2 - \lambda_3}{\lambda_2 + \lambda_3} \stackrel{\leq}{=} \frac{\lambda_1 - \lambda_3}{\lambda_1 + \lambda_3} \cdot \frac{\lambda_2 - \lambda_4}{\lambda_2 + \lambda_4},$$

we find

$$\sin\phi_{14} = \frac{4-1}{4+1} = \frac{3}{5}, \quad \phi_{14} = 36.87°,$$

$$\sin\phi_{23} = \frac{3-2}{3+2} = \frac{1}{5}, \quad \phi_{23} = 11.537°,$$

$$\sin \phi_{13} = \frac{3-1}{3+1} = \frac{1}{2}, \quad \phi_{13} = 30°,$$

$$\sin \phi_{24} = \frac{4-2}{4+2} = \frac{1}{3}, \quad \phi_{24} = 19.47°.$$

Putting these numerical values into the inequality, we have

$$\frac{3}{25} = 0.12 < \frac{1}{6} = 0.1667.$$

In Fig. 6.1(b) I sketched the relative locations of the corresponding angles. The inequality for the sine products would seem to imply some corresponding new trigonometry for the angles, e.g., a domination by the two

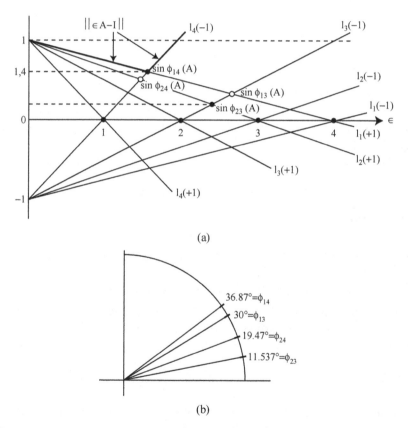

(a)

(b)

Fig. 6.1. New sine geometry of Hotelling inequalities.
(a) The four operator sines of the inequality.
(b) The corresponding angle relationships (not to scale).

inside angles taken together over the two outside angles. In this example, the inside angle sum does exceed the outside angle sum, but one cannot expect that to be the situation in general. Probably there is an interesting new sinusoidal trigonometry of matrix convex minima waiting further development.

6.5 Prediction Theory: Association Measures

Recently in [Gustafson (2010e)] I somewhat pushed my operator trigonometry to canonical correlations between the past and the future of a stationary stochastic process. I had known the Helson, Szego and Lax prediction theory of past and future from unrelated work I had done in the past. But here, it will be operator trigonometric.

Let us consider discrete parameter stationary stochasatic processes X_t, e.g., as in [Pourahmadi (2001)] for convenience. Then you can define the canonical correlation as follows. Let $X = (X_1, \ldots, X_m)$ and $Y = (Y_1, \ldots, Y_n)$ be random vectors. Let Σ_{11} be the $m \times m$ matrix $\mathrm{Cov}(X)$, Σ_{22} be the $n \times n$ matrix $\mathrm{Cov}(Y)$, and Σ_{12} the $m \times n$ matrix $\mathrm{Cov}(X, Y)$. Then the first canonical correlation is

$$\rho_1 = \max_{a,b} \frac{a' \Sigma_{12} b}{\sqrt{a' \Sigma_{11} a} \sqrt{b' \Sigma_{22} b}}.$$

Here a and b are vectors of coefficients which linearly combine the given random vectors to produce this largest canonical correlation. The succeeding $\rho_2, \ldots, \rho_m$ are found the same way with Rayleigh–Ritz orthogonality constraints relative to the previous subspaces.

The first point to note is that ρ_1 defines an angle by setting $\cos \phi = \rho_1$. Because ρ_1 is a maximum, this angle is a minimum. Please note that this distinguishes the canonical correlation theory from my antieigenvalue theory, where

$$\rho_1 = \cos \phi(A) = \min \frac{\langle Ax, x \rangle}{\|Ax\| \|x\|}$$

gives you a maximum angle. See Exercise 5 at the end of this chapter.

Let me now jump ahead to the angle between past and future. Let X_t be a discrete parameter stationary stochastic process, let $P = \overline{\mathrm{sp}}\{X_t, \ t \leq 0\}$ be the subspace of its past (and present) and let $F = \overline{\mathrm{sp}}\{X_t, \ t = 1, 2, \ldots\}$ be the subspace of its (strict) future. Let the past–future first (largest) canonical correlation be

$$\rho_1 = \sup\{|\text{Corr}\,(\mathbf{X}, \mathbf{Y}); \mathbf{X} \in P, \mathbf{Y} \in F|\}.$$

Introduce an angle θ according to $\cos\theta = \rho_1$ as above. One says that subspaces P and F are at a positive angle if $\theta > 0$, equivalently $\rho_1 < 1$. There is a large and technical Prediction Theory which elaborates these notions. Let me just mention that much of that theory can be helpfully focused by remembering the old F. and M. Riesz theory of Hardy spaces and the decomposition $L^2_+(S^1)$ and $L^2_-(S^1)$ of functions on the complex circle.

As pointed out in [Pourahmadi (2001, p. 294)], the paper [Jewell, Bloomfield and Bartmann (1983)] looked more closely at how to compute canonical correlations of past and future. They use the notation λ_1 instead of ρ_1 for the first canonical correlation but I stay with ρ_1 here. In their theory, ρ_1 is the largest eigenvalue of $H^*H = I - T^*T$ where H is a Hankel and T is a Toeplitz operator with symbol $\bar{\mathbf{h}}/\mathbf{h}$ where $\mathbf{h}$ is an outer function in the Hardy space H^2. If $\rho_1 < 1$, i.e., if the past and future subspaces are at a positive angle, then it is desirable to try to bound it away from 1, to get some estimation of the subspace angle θ. Using a spectral density argument they represent the symbol $\bar{\mathbf{h}}/\mathbf{h}$ in the form

$$\frac{\bar{\mathbf{h}}}{\mathbf{h}} = c e^{i(\mathbf{v} - \tilde{\mathbf{u}})}$$

where $|e| = 1$ and where $\tilde{\mathbf{u}}$ denotes the Hilbert transform of $\mathbf{u}$. Then their result is that ρ_1 is bounded above and below 1 according to

$$\rho_1 \overset{\le}{=} \frac{[1 + \|e^{\mathbf{u}}\|^2_\infty \|e^{-\mathbf{u}}\|^2_\infty - 2\|e^{\mathbf{u}}\|_\infty \|e^{-\mathbf{u}}\|_\infty \cos(2\|\mathbf{v}\|_\infty)]^{1/2}}{1 + \|e^{\mathbf{u}}\|_\infty \|e^{-\mathbf{u}}\|_\infty}.$$

This is a very technical result, but may be felt intuitively as the way the law of cosines enters into the computation of the Poisson kernel, e.g., see my book [Gustafson (1999f), p. 185].

Now I can connect to my operator trigonometry. In the special case that the general spectral density $w = \exp(\mathbf{u} + \mathbf{v})$ with $\mathbf{u} \in L^\infty$, or more generally when $m \overset{\le}{=} w \overset{\le}{=} M$, then the above bound simplifies to

$$\rho_1 \overset{\le}{=} \frac{M - m}{M + m}.$$

Since the "eigenvalue" of the joint dispersion matrix of the past and future are just the values of w, we may take this to mean that such matrix $\mathbf{A}$ is

SPD and the above bound becomes

$$\rho_1 \overset{\le}{=} \frac{\lambda_{\max} - \lambda_{\min}}{\lambda_{\max} + \lambda_{\min}} = \sin \phi(\mathbf{A}).$$

When equality holds, that means the largest canonical correlation $\rho_1 = \cos \theta = \sin \phi(\mathbf{A})$, thereby connecting to my operator trigonometry.

Jewel et al. (1983) go on to show how an alternate derivation yields a best constant k, which I elaborate here as

$$k = \frac{2Mm}{M + m} = \frac{2\sqrt{Mm}}{M + m} \cdot \sqrt{Mm} = (\lambda_{\max}\lambda_{\min})^{1/2} \cos \phi(\mathbf{A}).$$

Here I have gone ahead in this expression to factor to show the relation to the operator trigonometry. No doubt there are many more interesting future connections to be obtained here.

Now I turn to another recent paper. Canonical correlations are linked to statistical association measures (e.g., correlation coefficients) and to gaps between subspaces of a Hilbert space in [Dauxois, Nkiet and Romain (2004)]. In the [Dauxois et al. (2004)] paper, the lattice $\mathcal{F}(H)$ of closed subspaces of a real Hilbert space H is considered. Then relative orthogonal projections onto subspaces of the form $\overline{H_k + H_3} \ominus H_3$ are considered, where H_1, H_2 and H_3 are three selected subspaces and $k = 1$ or 2. The maximal canonical correlation coefficient between H_1 and H_2 relative to H_3 is defined as

$$\rho_1 = \rho_{123} = \text{spectral radius of } \hat{\mathbf{A}}$$

where $\hat{\mathbf{A}}$ is a positive selfadjoint operator defined in terms of the projections.

Here is my analysis from [Gustafson (2010e)]. Let $\mathbf{B}_{12} = (\mathbf{P}_{13} + \mathbf{P}_{23} - I)(\mathbf{P}_{13} + \mathbf{P}_{23})$ where the $\mathbf{P}_{k,3}$ are for example the above-described orthogonal projectors. Note that

$$\mathbf{B} = (\mathbf{P}_{13} + \mathbf{P}_{23})^2 - (\mathbf{P}_{13} + \mathbf{P}_{23})$$

$$= \mathbf{P}_{13}^2 + \mathbf{P}_{23}^2 + \mathbf{P}_{13}\mathbf{P}_{23} + \mathbf{P}_{23}\mathbf{P}_{13} - \mathbf{P}_{13} - \mathbf{P}_{23}$$

$$= \mathbf{P}_{13}\mathbf{P}_{23} + \mathbf{P}_{23}\mathbf{P}_{13}.$$

So far I have only used that the $\mathbf{P}$'s are (allowably even oblique) projectors. So $\mathbf{B}_{12}$ is the anticommutator $\text{Re}\,\mathbf{P}_{13}\mathbf{P}_{23}$. Next require the subspace projectors to be the orthogonal ones. Then $\mathbf{B}_{12}$ is selfadjoint. Let $\mathbf{A} \equiv \mathbf{A}_{12} = \mathbf{B}_{12}^2$. Then $\mathbf{A}$ is positive selfadjoint with spectral radius $\rho_1 \overset{\le}{=} 1$. If $\mathbf{A}$ has discrete spectrum, then its largest eigenvalue is $\lambda_1 = \rho_1$, and its smaller eigenvalues can be taken to be the successively smaller canonical correlation coefficients. If **A** is compact, then only 0 is in its continuous spectrum, so there again you will get discrete eigenvalue canonical correlations converging to zero.

The new link to my operator trigonometry is now apparent. If **A** has discrete spectra, you can use its (canonical correlation) eigenvalues to define my critical operator turning angles via the operator trigonometry $\sin \phi_i(\mathbf{A})$ formula.

Dauxois *et al.* (2004) go on to discuss subspace gaps and principal angles between subspaces. For further information on these, see [Kato (1976); Meyer (2000)]. Please remember that these principal subspace angles are minimum, not maximum, angles between subspaces. My operator trigonometry uses maximum turning angles. I have included further clarification of this distinction in Exercise 5.

6.6   Antieigenmatrices

In [Gustafson (1999e; 2002a)] I mentioned how numerous inequalities are already generalized in the literature to matrix rather than vector arguments. See for example the paper [Wang and Ip (2000)] which I mentioned there. Such leads one to wonder about a general theory of antieigenmatrices. One can easily define *ad hoc* several kinds of antieigenmatrices. What might those be?

To me, if they are to accord with my original concept of antieigenvectors as those most turned by an operator A, or most twisted by A if we take that perhaps better geometrical picture, then we would want that to be the defining concept. The issue would seem to be, what exactly do you want in a most twisted matrix? I have various notes and jottings about such a theory but nothing final. Probably we need a perfect application to guide us to a perfect definition.

One candidate of such application came before me when I saw the paper [Chaitin–Chatelin and Gratton (2000)]. As I discussed in Sec. 4.6, they factor an arbitrary real $n \times n$ matrix $A = QH$ into its polar representation and then investigate its absolute condition number $C(H)$. Their context is the Frobenius matrix norm $\|A\|_F^2 = \Sigma|a_{ij}|^2$. To illustrate my result that

$C^2(H) = 1 + \sin^2 \phi(H)$, I constructed the matrix

$$B = \begin{bmatrix} 0 & \cdots & \frac{\sigma_1}{(\sigma_1^2+\sigma_n^2)^{1/2}} \\ \frac{\sigma_n}{(\sigma_1^2+\sigma_n^2)^{1/2}} & \cdots & 0 \end{bmatrix}$$

to produce a maximal turning dynamics in the Frobenius inner product.

An easier definition arises just by lining up the antieigenvectors $x_\pm^i$ as columns in the regressor matrix X to produce worst efficiency in the DBWK bound.

But then for every different matrix inequality you might want a different definition of antieigenmatrix. Such inequalities generalized from vectors to matrices go back to [Marshall and Olkin (1990)] and even further to [Fan (1966)]. So for now I have no definitive definition of antieigenmatrices. However, in Chapter 7 we will see some compelling evidence for what they should do within a context of quantum mechanics.

Commentary

In [Gustafson (1972a)] I took terse note of possible future applications of my new antieigenvalue theory to statistics, with the statement ". . . it should be pointed out that μ_1 is the ratio of moments of A and it might be fruitful to examine μ_1 from the viewpoint of moment theory. . . ." There I was viewing my antieigenvalue variational functional $\mu(x)$ in a loose variational-to-moment interpretation

$$\mu(x) = \frac{\langle Ax, x \rangle}{\langle A^2x, x \rangle^{1/2}\langle x, x \rangle^{1/2}} = \frac{\mu_1}{\mu_2^{1/2}\mu_0^{1/2}}.$$

However I did not follow up at that time. Now we see that my intuition was actually anticipating the situation of canonical correlations in statistics as treated later by C. R. Rao and C. V. Rao (1987) who looked at Rayleigh quotients of the form

$$\frac{\langle Cx, x \rangle}{\langle Ax, x \rangle^{1/2}\langle Bx, x \rangle^{1/2}},$$

which I discussed in Sec. 6.3.

In that section I also delved briefly into the electrical engineering control application (1980) which led the Rao's to consider such Rayleigh quotients for application to matrix statistics. I could have put that into the short control theory Sec. 5.5. But it is more fruitfully discussed here.

The DBWK conjectured lower bound of Sec. 6.2 remained an open problem for twenty years after Durbin conjectured it. See the history given in [Bloomfield and Watson (1975)].

As I pointed out in [Gustafson (2002a)], Bloomfield and Watson actually drifted rather close to my operator trigonometry when they stated in (1975), p. 126, using my notation here rather than their notation:

"*In (2.8) we are interested in the relationship of each term in the product and unity. But*

$$\frac{(\lambda_\ell + \lambda_{n-\ell+1})^2}{4\lambda_\ell \lambda_{n-\ell+1}} - 1 = \frac{(\lambda_\ell - \lambda_{n-\ell+1})^2}{4\lambda_\ell \lambda_{n-\ell+1}} ."$$

If you divide everything by the first term you arrive at

$$1 - \frac{4\lambda_\ell \lambda_{n-\ell+1}}{(\lambda_\ell + \lambda_{n-\ell+1})^2} = \left(\frac{\lambda_\ell - \lambda_{n-\ell+1}}{\lambda_\ell + \lambda_{n-\ell+1}}\right)^2 .$$

This my operator trigonometry recognizes as

$$1 - \cos^2 \phi_{\ell,n-\ell+1}(A) = \sin^2 \phi_{\ell,n-\ell+1}(A).$$

But one needs my Min-Max Theorem to see that.

As I mentioned in the *Perspective* to this chapter, my breakthrough paper [Gustafson (1999e)] was rejected by one referee when I submitted that paper in 1999 to a well-known statistics journal. But I knew my connection of my operator trigonometry to statistics was important, and mentioned it in [Gustafson (2000d; 2001a)] and it is implicit in [Gustafson (1999b)]. I am indebted to Simo Puntanen and George Styan, who welcomed me into their matrix statistics community and invited me to present my results at their series of IWMS conferences.

Subsequently, [Khatree (2001; 2002; 2003)] extended some of my results, and provided some further statistics context. Then C. R. Rao (2005; 2007) gave his versions of my antieigenvalue theory as applied to matrix statistics. I have a few differences with their treatments, and you may find those explained in [Gustafson (2006b)]. A strong point of C. R. Rao's treatment is that it extends my results to the wider and useful context permitting generalized inverses.

My trigonometric proof of the Hotelling inequalities in Sec. 6.4 started in [Gustafson (2006b)] and I finished the last piece of it for the general case in [Gustafson (2009b)]. While doing so I searched through the old literature and in particular papers by Hotelling and never found that inequality. However, the name Hotelling was attached to it in [Drury *et al.* (2002)], and that surely is a good name for a nice inequality.

It was from a discussion with Mohsen Pourahmadi at the IWFOS 2008 conference in Toulouse, France, see my short summary [Gustafson (2008a)] of my lecture there, that I became familiar with his book [Pourahmadi (2001)], which provided me ready access to prediction theory in a way in which I could implement aspects of my operator trigonometry, as presented in Sec. 6.5.

I first mentioned the concept of antieigenmatrices in the preprint [Gustafson (1999e)]. However, as I have made clear in Sec. 6.6, there were already precursors for such earlier in the matrix statistics inequalities literature.

6.7   Exercises

1. (History) Explore further the rather unexpected but quite interesting historical happening this author uncovered: that both the theory of statistical inefficiency following from the work of John Von Neumann, and the nonanticipatory stochastic calculus developed by Kiyoshi Ito originated at least partly from their World War II work on ballistics.
2. Verify Von Neumann's eigenvalues mentioned in Sec. 6.1. Develop further the proposed synergistic relationship between statistical efficiency and partial differential equation boundary value problems.
3. That the inefficiency equation, and in the same way the Euler equation, possesses only two (or one) component solutions, follows by a reducing subspace argument. It goes as follows. Multiply the Euler equations by A^{-1} to obtain

$$\frac{Ax}{\langle A^2 x, x \rangle} - \frac{2x}{\langle Ax, x \rangle} + A^{-1}x = 0.$$

Then notice that any vector w in the span sp $\{x, Ax\}$ for any solution x of this Euler equation is mapped into this span by A. I traced this argument all the way back to the original paper of Bloomfield and Watson (1975). Work through the same details for the inefficiency equation, as they did in 1975.

4. Look further into possible interesting operator trigonometry that can be worked out for the electrical engineering applications discussed in Sec. 6.3.

5. As emphasized in Sec. 6.4 and earlier in this book, my antieigenvalue theory and operator trigonometry conceptualized the notion of a maximally turned vector, which I called the first antieigenvector. This concept should be distinguished from the well-known theory of principle angles between subspaces, e.g., see [Meyer (2000); Kato (1976)]. Because of confusion I sometimes encounter as concerns this distinction, I sometimes quickly make the explanatory point that my antieigenvectors are perhaps better seen geometrically as the maximally twisted ones. I kill two birds with one stone with that explanation. The twist is not related to principle angles. The other bird is the idea of rotation, e.g., as in rotation matrices, or as in the unitary part of an operator $A = U|A|$ when expressed in polar factorization. It is the twisting in $|A|$ that is being measured, not the group actions residual in U.

In this exercise I would like to further clarify this distinction of my theory from that of subspace principal angles. To do so, I will make use of the nice simple presentations of subspace principal angle theory that you will find in [Meyer (2000)], to which I refer the reader. For clarity I consider here only the finite-dimensional vector space situation, even just the R^n case. When $R^n = R \oplus N$ the subspace angle θ between R and N is defined to be that of

$$\cos \theta = \max_{\substack{\mathbf{u} \in R \\ \mathbf{v} \in N}} \frac{\mathbf{v}^T \mathbf{u}}{\|\mathbf{v}\|_2 \|\mathbf{u}\|_2} = \max_{\substack{\mathbf{u} \in R, \mathbf{v} \in N, \|\mathbf{u}\|_2 = \|\mathbf{v}\|_2 = 1}} \mathbf{v}^T \mathbf{u}.$$

Let $\mathbf{P}$ be the (generally oblique) projector of R^n onto R along N. $\mathbf{P}$ has two-norm

$$\|\mathbf{P}\|_2 = \max_{\|x\|_2 = 1} \|\mathbf{P}\mathbf{x}\|_2 = \sqrt{\lambda_{\max}(\mathbf{P}^T \mathbf{P})} = \frac{1}{\sin \theta}$$

where θ is the above angle between the two subspaces.

Now let us turn to my operator angle

$$\cos \phi(A) = \min_{x \neq 0} \left\langle \frac{\mathbf{A}\mathbf{x}}{\|\mathbf{A}\mathbf{x}\|}, \frac{\mathbf{x}}{\|\mathbf{x}\|} \right\rangle,$$

where I have written it in such a way to emphasize that it measures a maximum turning by $\mathbf{A}$ of vectors normalized on the unit sphere. So we see that we are just looking at one-dimensional subspaces sp {x} and sp {Ax} but we are letting them run over all such subspace pairs.

Let us now force a connection to the subspace angle context. Fix x, rule out eigenvectors, so in the two-dimensional space $M = \text{sp}\{x, Ax\}$, we may take the oblique projector $\mathbf{P_x}$ of sp{Ax} along sp{x}. Then we have an x-dependent angle $\boldsymbol{\theta_x}$ according to the above angle between subspaces theory:

$$\cos \theta_x = \max_{\|x\|=\|Ax\|=1} \langle \mathbf{Ax}, \mathbf{x} \rangle = \frac{\langle \mathbf{Ax}, \mathbf{x} \rangle}{\|\mathbf{Ax}\|\, \|\mathbf{x}\|}.$$

But since x is fixed, there is really no maximum or minimum involved. My operator trigonometry angle $\phi(\mathbf{A})$ then takes the maximum of these angles θ_x, i.e., the minimum of $\cos \theta_x$, as x runs over all directions. To firm up your understanding of my theory, prove that in the case of eigenspaces of multiplicity greater than one, in my antieigenvector expression

$$x_{\pm} = \pm \left(\frac{\lambda_n}{\lambda_1 + \lambda_n} \right)^{1/2} x_1 + \left(\frac{\lambda_1}{\lambda_1 + \lambda_n} \right)^{1/2} x_n$$

you may take any x_1 of norm one from that eigenspace and similarly any x_n of norm one from that eigenspace and still produce the same maximal turning (twisting).

6. Oblique projectors were mentioned above in Sec. 6.5 and in the above Exercise 5. Already in Exercise 3 of Chapter 1 we showed that every $n \times n$ matrix A of rank k has a spectral representation

$$A = c_1 P_1 + \cdots + c_k P_k$$

where the P_i are oblique projectors. Within a context of finite matrices, this nicely contains all of the other spectral representation theorems for selfadjoint and normal operators, the singular value decomposition, and so on. Moreover, it allows great freedom in choosing whatever projectors you may want. It is my feeling that as one moves beyond symmetric matrices within the domain of statistics and also elsewhere, more attention will be given to the role of oblique projectors.

Stated another way, as one moves beyond normal matrices, there is no implicit expectation to view them as naturally constituted from orthogonal projections. Granted, orthogonal projections and orthogonal bases do have rule-of-thumb advantages, e.g., minimizing round-off errors.

On the other hand, oblique projections abound. Show that in any linear space X and for any subspaces M and N such that $M \cap N = \{0\}$ and $X = M+N$, an oblique projector P results from writing each x as $x = m \oplus n$ and defining P according to: $Px = m$. What is the geometry of this projector? Does it have a trigonometry?

Chapter 7

Quantum Trigonometry

Perspective

The opportunity to apply my operator trigonometry to quantum mechanics in a productive way came about by chance at a luncheon at the elegant Schlumberger estate Les Treilles in Provence in July 1996. The story is worth telling for two reasons. Quantum mechanics is a very important thing which runs the universe. And much of scientific discovery depends upon the web of human relationships. Both depend fundamentally on the elements of chance.

Before I tell the story, of course be sure that I had been touching quantum mechanics in various ways for thirty years before 1996, but more from the points of view of Schrodinger's partial differential equation, quantum scattering theory, and especially the baffling mysteries of the quantum measurement problem. But I had never looked at Bell's inequalities.

Through my scientific friendship with Nobel Laureate Ilya Prigogine and my role as an honorary member of his Solvay Institute for physics in Brussels, I had the great fortune to be invited to the prestigious Fondation des Treilles estate nestled deep in Provence for three scientific retreats, in 1994, 1996, and 1999. One would arrive, be given a private car to drive to your private villa, with only the expectation that you show up at 9:00 each morning for intensive but relaxed discussions all day on the current pressing issues in physics.

At the 1996 meeting, Luigi Accardi, a brilliant Italian who specializes in quantum probability, felt like giving a rather expansive after-lunch informal presentation of some of his latest results. At one point it all came down to a certain 2×2 matrix. Luigi started to apologize for such a small matrix, but I encouraged him by commenting that I had found 2×2

matrix products not to be underestimated, as concerns for example, loss of positivity.

A year and a half later, on January 27, 1998, out of the blue came a letter from Professor Accardi, saying that he wanted to know more about precisely the point of my remark, and that he was sending me a number of his papers on quantum probability. Although those papers (very numerous...) did not make direct contact to my operator trigonometry, from them I tracked down the [Accardi and Fedullo (1982)] paper.

This paper brought me to Bell's inequalities. First, Accardi and Fedullo show that one has a Kolmogorovian probability model iff certain conditional probabilities p, q, and r satisfy the inequality

$$|p + q - 1| \overset{\leq}{=} r \overset{\leq}{=} 1 - |p - q|.$$

Then in terms of two-dimensional complex Hilbert spaces they showed that a quantum probabilistic spin model exists iff there exist three corresponding normalized three-vectors a, b, and c such that

$$\cos^2 \alpha + \cos^2 \beta + \cos^2 \gamma - 1 \overset{\leq}{=} 2 \cos \alpha \cos \beta \cos \gamma,$$

where the angles α, β, and γ are the angles between the sought-for directions a, b, and c. But that inequality is equivalent to the triangle inequality ($\ast\ast\ast$) of my operator trigonometry, see Sec. 3.5.

I immediately could see how to put my operator trigonometry into quantum mechanics. There was some work to do of course. As a consequence, my geometrical trigonometrization of the Bell theory provides considerable mathematical and physical clarification of critical foundational issues in quantum mechanics. These results will be presented in Sec. 7.1. In Sec. 7.2 we go further and generalize the Bell-type inequalities to what I call quantum spin-trigonometry identities. These explain the Bell violation zones in purely mathematical terms.

I have followed the attempts to develop quantum computers now for about twenty years. In 2000 I was asked to plenary at such a conference and I took that opportunity to try to bring my operator trigonometry into that arena. I am not satisfied with my theory as it pertains to unitary operators — the setting of quantum computer models — but I present my approach here in Sec. 7.3 anyway.

Occasioned by a week's visit by R. Penrose and his lectures for us at the end of January 2007, and several nice conversations I had with him, it occurred to me one day during that week to work out the essentials of my operator trigonometry for the basic twistor matrices which appeared in his lectures so that I could quickly discuss such with him while he was in Boulder. Those results will be presented in Sec. 7.4. Then in Sec. 7.5 I go further and look at elementary particles. Finally, in Sec. 7.6 I give my very recent development of the trigonometry of quantum states (i.e., density matrices).

7.1   Bell–Wigner–CHSH Inequalities

The famous EPR paper of [Einstein, Podolski and Rosen (1935)] raised serious issues about the completeness of the quantum mechanical formalism — issues that are unresolved to this day. Approximately thirty years later the equally famous paper of [Bell (1964)] appeared, in which he presented his Bell's inequality

$$|P(\mathbf{a}, \mathbf{b}) - P(\mathbf{a}, \mathbf{c})| \overset{\le}{=} 1 + P(\mathbf{b}, \mathbf{c})$$

and exhibited quantum spin measurement configurations whose quantum expectation values could not satisfy his inequality. Bell's analysis assumed that two measuring apparatuses could be regarded as physically totally separated, and free from any effects of the other. Thus his inequality could provide a test which could be failed by measurements performed on correlated quantum systems. It was therefore argued that local realistic hidden variable theories could not hold in quantum mechanics if future physical experiments would violate Bell's inequality. Later, Aspect, Dalibard, and Roger (1982) indeed demonstrated violation of Bell's inequality in their laboratory experiments. But the controversy about the Bell inequality and related inequalities to be mentioned below and their physical implications continues to this day.

Bell's arguments in arriving at his inequality were classical probabilistic correlation arguments. Thereupon Wigner (1970) presented his own version, making clearer the issues of locality and so-called realism. Furthermore, Wigner was sure to use quantum mechanical probabilistic correlations. His version of Bell's theory then becomes the inequality

$$\frac{1}{2}\sin^2\frac{1}{2}\theta_{23} + \frac{1}{2}\sin^2\frac{1}{2}\theta_{12} \overset{\ge}{=} \frac{1}{2}\sin^2\frac{1}{2}\theta_{31}$$

where the θ_{ij} are angles between spin direction ω_i and ω_k. See Fig. 7.1(a).

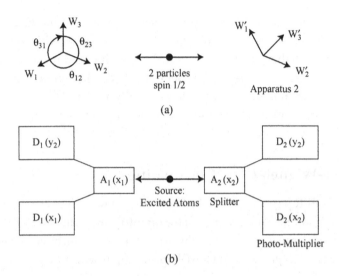

(a)

(b)

Fig. 7.1. Experimental set-ups for Bell inequalities.
(a) Nine measurements to simultaneously measure the direction vectors of the two spins.
(b) Source emits statistical ensemble of polarization-correlated photon couples. Polarizer A decomposes the incident light into ordinary ray polarized along x_1 direction and extraordinary polarized along y_1 perpendicular direction for measurement at detectors D.

Another important Bell-type inequality is that of [Clauser, Horne, Shimony, and Holt (1969)]. Let $\mathbf{a}, \mathbf{b}, \mathbf{c}, \mathbf{d}$ be four arbitrary chosen unit vector directions in plane orthogonal to the two beams produced by the source. Let $v_i(\mathbf{a})$ and $v_i(\mathbf{d})$ be the "hidden" predetermined values ± 1 of the spin components along $\mathbf{a}$ and $\mathbf{d}$, respectively, of particle 1 of the ith pair; similarly $w_i(\mathbf{b})$ and $w_i(\mathbf{c})$ for particle 2 values along directions $\mathbf{b}$ and $\mathbf{c}$. Then the average correlation value for particle 1 spins measured along $\mathbf{a}$ and particle 2 spins measured along $\mathbf{b}$ is $E(\mathbf{a}, \mathbf{b}) = \sum_{i=1}^{N} v_i(\mathbf{a})w_i(\mathbf{b})/N$. Taking into account in the same way the average correlation values $E(\mathbf{a}, \mathbf{c})$, $E(\mathbf{d}, \mathbf{b})$, $E(\mathbf{d}, \mathbf{c})$ and adding up all pairs, one arrives at the CHSH inequality

$$|E(\mathbf{a}, \mathbf{b}) + E(\mathbf{a}, \mathbf{c}) + E(\mathbf{d}, \mathbf{b}) - E(\mathbf{d}, \mathbf{c})| \stackrel{\leq}{=} 2.$$

In a series of papers starting with [Gustafson (1999a), see also (2000a,c; 2001b; 2002b; 2003b–d; 2005c; 2006a; 2007a)], I placed all of the inequalities of the Bell theory, into my noncommutative trigonometry. From my results, it may be argued that many important issues in the Bell theory, about which there are at times furious arguments about physical and probabilistic meaning among physicists, are really better seen as new mathematical quantum geometry from the noncommutative trigonometry.

I cannot do justice here to all the important physical matters which one must take into account when working conscientiously on these issues and finely tuned brilliant experiments attempting to resolve them. You will find some further discussion in my papers cited above, and as well important key references from the physics literature. A good somewhat recent book is that of [Afriat and Selleri (1999)]. Many different quantum particles beyond Bell's gedanken-imagined spin 1/2 particles have been treated both theoretically and experimentally. In Fig. 7.1(b) I have schematically drawn one such general experimental set-up.

For example as explained in [Gustafson (2000c)], Wigner (1970) made a number of assumptions, among which are: all possible measurements are predetermined, any measurement on one apparatus cannot change the preset outcomes on the other apparatuses, etc. Then two spin-1/2 particles coming from a common source, with perfect anticorrelation and singlet properties, are sent to the two detectors. Nine measurements are then needed to measure the direction vectors $\omega_1, \omega_2, \omega_3$ of the two spins. Each spin has two possible values: $1/2 \equiv +, -1/2 \equiv -$, so each measurement can permit four relative results: $++, --, +-, -+$. Thus there are 4^9 possible outcomes. Wigner then assumes that the spins are not affected by the orientation of the particular measuring apparatus. This reduces the outcomes to 2^6 possibilities. He then further reduces the possibilities to four possible cases, each of which possesses a corresponding inequality for experimental testing. The Wigner inequality I gave above corresponds to the $++$ outcome with the first particle in direction ω_1 and the second particle in direction ω_3. Then, assuming a joint probability factorization, an assumption made also by Bell and to be sure a source of contention thereafter but which need not concern us here, Wigner states that the "hidden parameters can reproduce the quantum mechanical probabilities only if the three directions $\omega_1, \omega_2, \omega_3$ in which the spins are measured are so situated that $1/2 \sin^2 \theta_{23}/2 + 1/2 \sin^2 \theta_{12}/2 \overset{\geq}{=} 1/2 \sin^2 \theta_{31}/2$," i.e., the inequality above. Then to make the point very clear, he specializes to the case in which the three directions $\omega_1, \omega_2, \omega_3$ in three-space are coplanar and with ω_2 bisecting the angle between ω_1 and ω_3. Then $\theta_{12} = \theta_{23} = \theta_{31}/2$ and inequality $++$ becomes

$$\sin^2\left(\frac{1}{2}\theta_{12}\right) \geq \frac{1}{2}\sin^2(\theta_{12}) = 2\sin^2\left(\frac{1}{2}\theta_{12}\right)\cos^2\left(\frac{1}{2}\theta_{12}\right)$$

from which $\cos^2(\theta_{12}/2) \leq 1/2$ and hence $\theta_{31} = 2\theta_{12} \geq \pi$. Thus the inequality necessary for appropriate quantum mechanical spin probabilities for the hidden variable theories is violated for all $\theta_{31} < \pi$. Wigner then asserts (without giving the details) that the same conclusion may be drawn for all coplanar directions.

Let us now look at this conclusion and its extension to all coplanar directions from the operator trigonometric perspective. The Gram determinant G of Sec. 3.5 vanishes if and only if the three directions are coplanar, no matter what their frame of reference. Then we may write the general triangle inequality of Sec. 3.5 as follows:

$$(1 - a_1^2) + (1 - a_2^2) - (1 - a_3^2) = 2a_3(a_3 - a_1a_2)$$

or in the terminology of Wigner's quantum spin situation, as

$$\sin^2\left(\frac{1}{2}\theta_{12}\right) + \sin^2\left(\frac{1}{2}\theta_{23}\right) - \sin^2\left(\frac{1}{2}\theta_{13}\right)$$
$$= 2\cos\left(\frac{1}{2}\theta_{13}\right)\left[\cos\left(\frac{1}{2}\theta_{13}\right) - \cos\left(\frac{1}{2}\theta_{12}\right)\cos\left(\frac{1}{2}\theta_{23}\right)\right].$$

Nonquantum probability violation in the coplanar case is equivalent to the right-hand side being nonnegative. Since all half-angles do not exceed $\pi/2$, except for the trivial case when $\theta_{13}/2 = \pi/2$, the nonnegativity means that of its second factors. By choosing the direction ω_2 to be the "one in-between" among the half-angles, we can without loss of generality assume that $\theta_{12}/2 + \theta_{23}/2 = \theta_{13}/2$. The required nonnegativity then reduces by the elementary cosine sum formula to

$$\cos\left(\frac{\theta_{12} + \theta_{23}}{2}\right) \geq \frac{1}{2}\left[\cos\left(\frac{\theta_{12} + \theta_{23}}{2}\right) + \cos\left(\frac{\theta_{12} - \theta_{23}}{2}\right)\right],$$

i.e., $\cos((\theta_{12} + \theta_{23})/2) \geq \cos((\theta_{12} - \theta_{23})/2)$, which is false for positive θ_{23}. This completes Wigner's argument and is the meaning of coplanar quantum probability violation.

Wigner considers two other configurations for quantum nonprobability violation, namely

$$1 - \frac{1}{2}\left(\sin^2 \frac{1}{2}\theta_{12} + \sin^2 \frac{1}{2}\theta_{23} + \sin^2 \frac{1}{2}\theta_{31}\right) \overset{\geq}{=} 0$$

and

$$\frac{1}{2}\left(\sin^2 \frac{1}{2}\theta_{12} + \sin^2 \frac{1}{2}\theta_{23} - \sin^2 \frac{1}{2}\theta_{31}\right) \overset{\geq}{=} 0.$$

As Wigner notes, the positivity of the three cyclically interchanged versions of these is a necessary and sufficient condition for the possibility to interpret the spin measurements in the ω_i directions on a singlet state in terms of hidden variables.

How do these look from the operator trigonometric perspective, using only the Gram matrix and cosine sums? Using the general Gramian expression

$$|G| = (1 - a_1^2) + (1 - a_2^2) + (1 - a_3^2) + 2(a_1 a_2 a_3 - 1)$$

immediately the two inequalities above become, respectively

$$a_1 a_2 a_3 - \frac{1}{2}|G| \overset{\geq}{=} 0$$

and

$$2a_3(a_1 a_2 - a_3) \overset{\geq}{=} 0.$$

Thus the operator trigonometry perspective makes more precise both qualitatively and quantitatively the arguments of Wigner. In particular the fundamental triangle inequality and the Gramian of Sec. 3.5 are useful new ingredients in connecting the quantum spin probabilities to the operator trigonometry.

7.2 Trigonometric Quantum Spin Identities

Then I decided to go further and try to render these inequalities into equalities. An analogy is the putting into linear programming algorithms the so-called *slack variables*. But here my *inequality–equalities*, as I called them, would clearly show how certain previously unaccounted-for quantum probabilities will necessarily show up in the violation regions of the Bell theory.

As a first example, consider Wigner's version of the Bell inequality discussed above. For the three coplanar directions, our corresponding inequality equality becomes in Wigner's quantum spin setting terminology

$$\sin^2\left(\frac{1}{2}\theta_{12}\right) + \sin^2\left(\frac{1}{2}\theta_{23}\right) - \sin^2\left(\frac{1}{2}\theta_{13}\right)$$

$$= 2\cos\left(\frac{1}{2}\theta_{13}\right)\left[\cos\left(\frac{1}{2}\theta_{13}\right) - \cos\left(\frac{1}{2}\theta_{12}\right)\cos\left(\frac{1}{2}\theta_{23}\right)\right].$$

Violation of the conventionally assumed quantum probability rule $|\langle u, v \rangle|^2 \equiv \cos^2 \theta_{u,v}$ for unit vectors u and v representing prepared state u to be measured as state v, is equivalent according to Wigner to the right-hand side of this identity being negative. This is his Bell "violation" test. However, from our point of view, there is no violation, there is just a quantum trigonometric identity, valid for certain formulations of measurement of certain spin systems.

As a second example, let us consider the important CHSH inequality given above. Wishing now to preserve equality, we write

$$|\mathbf{a} \cdot \mathbf{b} + \mathbf{a} \cdot \mathbf{c} + \mathbf{d} \cdot \mathbf{b} - \mathbf{d} \cdot \mathbf{c}| = |\mathbf{a} \cdot (\mathbf{b} + \mathbf{c}) + \mathbf{d} \cdot (\mathbf{b} - \mathbf{d})|$$

$$= |\|b + c\| \cos \theta_{a,b+c} + \|b - c\| \cos \theta_{d,b-c}|$$

$$= |(2 + 2\cos\theta_{bc})^{1/2} \cos\theta_{a,b+c}$$

$$+ (2 - 2\cos\theta_{bc})^{1/2} \cos\theta_{d,b-c}|.$$

Squaring this expression and writing everything quantum-trigonometrically,

$$|\mathbf{a} \cdot \mathbf{b} + \mathbf{a} \cdot \mathbf{c} + \mathbf{d} \cdot \mathbf{b} - \mathbf{d} \cdot \mathbf{c}|^2 = (2 + 2\cos\theta_{bc})\cos^2\theta_{a,b+c}$$

$$+ (2 - 2\cos\theta_{bc})\cos^2\theta_{d,b-c}$$

$$+ 2(4 - 4\cos^2\theta_{bc})^{1/2}\cos\theta_{a,b+c}\cos\theta_{d,b-c}$$

$$= 4\cos^2\left(\frac{\theta_{bc}}{2}\right)\cos^2\theta_{a,b+c}$$

$$+ 4\sin^2\left(\frac{\theta_{bc}}{2}\right)\cos^2\theta_{d,b-c}$$

$$+ 4\sin\theta_{bc}\cos\theta_{a,b+c}\cos\theta_{d,b-c}.$$

In the above we used two standard trigonometric half-angle formulas. Now substituting the double-angle formula $\sin \theta_{bc} = 2 \sin (\theta_{bc}/2) \cos (\theta_{bc}/2)$ into the above we arrive at

$$|\mathbf{a} \cdot \mathbf{b} + \mathbf{a} \cdot \mathbf{c} + \mathbf{d} \cdot \mathbf{b} - \mathbf{d} \cdot \mathbf{c}|^2 = 4 \left[\cos \left(\frac{\theta_{bc}}{2} \right) \cos \theta_{a,b+c} \right. $$
$$\left. + \sin \left(\frac{\theta_{bc}}{2} \right) \cos \theta_{d,b-c} \right]^2$$

and hence we have a new quantum *CHSH equality*

$$|\mathbf{a} \cdot \mathbf{b} + \mathbf{a} \cdot \mathbf{c} + \mathbf{d} \cdot \mathbf{b} - \mathbf{d} \cdot \mathbf{c}| = 2 \left| \cos \left(\frac{\theta_{bc}}{2} \right) \cos \theta_{a,b+c} + \sin \left(\frac{\theta_{bc}}{2} \right) \cos \theta_{d,b-c} \right|.$$

We may also write the right-hand side as twice the absolute value of the two-vector inner product

$$\mathbf{u}_1 \cdot \mathbf{u}_2 \equiv \left(\cos \left(\frac{\theta_{bc}}{2} \right), \sin \left(\frac{\theta_{bc}}{2} \right) \right) \cdot \left(\cos \theta_{a,b+c}, \cos \theta_{d,b-c} \right)$$

to arrive at the quantum trigonometric identity

$$|\mathbf{a} \cdot \mathbf{b} + \mathbf{a} \cdot \mathbf{c} + \mathbf{d} \cdot \mathbf{b} - \mathbf{d} \cdot \mathbf{c}| = 2(\cos^2 \theta_{a,b+c} + \cos^2 \theta_{d,b-c})^{1/2} |\cos \theta_{u_1,u_2}|.$$

The right-hand sides of these equalities isolate the "classical limiting probability factor" 2 from the second factor, which may achieve its maximum $\sqrt{2}$. Fix any directions $\mathbf{b}$ and $\mathbf{c}$. Then choose $\mathbf{a}$ relative to $\mathbf{b} + \mathbf{c}$ and choose $\mathbf{d}$ relative to $\mathbf{b} - \mathbf{c}$ so that $\cos^2 \theta_{a,b+d} = 1$ and $\cos^2 \theta_{d,b-c} = 1$, respectively. Now we may choose the free directions $\mathbf{b}$ and $\mathbf{c}$ to maximize the third factor to $\cos \theta_{u_1,u_2} = \pm 1$. But that means the two-vectors $\mathbf{u}_1$ and $\mathbf{u}_2$ are collinear and hence

$$\mathbf{u}_1 = \left(\cos \left(\frac{\theta_{bc}}{2} \right), \sin \left(\frac{\theta_{bc}}{2} \right) \right) = 2^{-1/2} (\cos \theta_{a,b+c}, \cos \theta_{d,b-c})$$
$$= 2^{-1/2} (\pm 1, \pm 1),$$

and thus the important angle θ_{bc} is seen to be $\pm \pi/2$. More to the point, our identity allows one to exactly trace out the "violation regions" analytically in terms of the trigonometric inner product condition $1 \overset{\leq}{=} |\mathbf{u}_1 \cdot \mathbf{u}_2| \overset{\leq}{=} \sqrt{2}$.

Thus, and we have only presented two examples here, my noncommutative trigonometry applied to the celebrated inequalities of the Bell theory has enabled much clearer delineations of the physical, probabilistic, and geometrical implications of those inequalities. In particular, in my opinion, one cannot claim quantum locality or nonlocality on the basis of satisfaction or violation of Bell inequalities.

7.3   Quantum Computing: Phase Issues

In [Gustafson (2001a)] I had the opportunity to insert a little of my operator trigonometry into quantum computing. The results are very preliminary and admittedly meager. Two good books on quantum computing are [Nielsen and Chuang (2000); Mermin (2007)]. I have no intention here of more than just scratching the surface of this audacious enterprise. To summarize: the problem is hardware. Then the software. The mathematics to make it go is the easiest part! That is to say, adequate theory and algorithms have been established to justify the devilish problem of keeping atoms or ions or whatever under control long enough to quantum compute. The range of hardware design runs from ion traps to solid state to polymers. These currently work but are not yet scalable up to any useful size.

Turning then to quantum computing and the potential use of the operator trigonometry therein, the main idea is to develop a trigonometric theory of decoherence and also perhaps for the dynamics of certain quantum computing algorithms. However, this will require a further extension of the operator trigonometry, because thus far I have extended in [Gustafson (2000d)] the theory from A symmetric positive definite to arbitrary $A = U|A|$ via polar form, at the expense of ignoring U. It is exactly unitary matrices we need to examine in quantum computing.

Let me illustrate the situation as follows. Early in the operator trigonometry I defined a "total" turning angle $\phi_{\text{total}}(A)$ for arbitrary operators A by

$$\cos \phi_{\text{total}}(A) = \inf_{Ax \neq 0} \frac{|\langle Ax, x \rangle|}{\|Ax\| \|x\|}.$$

For invertible normal operators A it is known, see e.g., the book [Gustafson (1997a)], the general formula for total antieigenvalues of normal operators, namely,

$$\cos^2 \phi_{\text{tot}}(A) = |\mu_1|^2(A) = \frac{(\beta_i|\lambda_j| + \beta_j|\lambda_i|)^2 + (\delta_i|\lambda_j| + \delta_j|\lambda_i|)^2}{(|\lambda_i| + |\lambda_j|)^2 |\lambda_i||\lambda_j|}.$$

There $\lambda_i = \beta_i + i\delta_i$ and $\lambda_j = \beta_j + i\delta_j$ are the two eigenvalues of a normal operator A which contribute to the first total antieigenvectors of A. For the matrix example

$$A = \begin{bmatrix} \sqrt{3}/2 + 1/2 & 0 \\ 0 & 1/2 + i\sqrt{3}/2 \end{bmatrix}$$

this means that

$$\cos \phi_{\text{tot}}(A) = \left[\frac{(\sqrt{3}/2 + 1/2)^2 + (1/2 + \sqrt{3}/2)^2}{(1+1)^2(1)(1)} \right]^{1/2} \approx 0.96592582.$$

We note that $\phi_{\text{tot}}(A) = 15°$ and that this is the angle of bisection of the angle between the eigenvalues λ_1 and λ_2. This simple unitary operator A indicates that we do not get much information from the total angle operator trigonometry.

I took this example from [Gustafson (2000d)] where it is shown that generally the three angles $\phi(A)$, $\phi_{\text{tot}}(A)$, and $\phi_e(A)$ may be different, the latter deriving meaning from the angle of A given by $\phi(|A|)$ in the extended operator trigonometry. Recall that the original $\phi(A)$ of the operator trigonometry was defined for use with accretive operators A and hence it is determined in terms of a "real" cosine. For A above we have

$$\frac{1}{2} = \cos \phi(A) < 0.966 = \cos \phi_{\text{tot}}(A) < 1 = \cos \phi_e(A).$$

Consider for example the quantum computing conditional phase shift operator

$$B(\phi) = \begin{bmatrix} 1 & 0 & 0 & 0 \\ 0 & 1 & 0 & 0 \\ 0 & 0 & 1 & 0 \\ 0 & 0 & 0 & e^{i\phi} \end{bmatrix}.$$

The total operator trigonometry turning angle $\phi_{\text{total}}(B(\phi))$ will be determined by the eigenvalues 1 and $e^{i\phi}$, from which we obtain

$$\cos^2 \phi_{\text{total}}(B(\phi)) = \frac{(\cos \phi + 1)^2 + (\sin \phi)^2}{4} = \frac{1 + \cos \phi}{2}.$$

Thus for example for phase $\phi = 45°$ we find $\phi_{\text{total}}(B(\phi)) = 22.5°$. As a second example we may consider the quantum controlled-not operator

$$C_{\text{XOR}} = \begin{bmatrix} 1 & 0 & 0 & 0 \\ 0 & 1 & 0 & 0 \\ 0 & 0 & 0 & 1 \\ 0 & 0 & 1 & 0 \end{bmatrix}.$$

The total antieigenvectors are determined by the eigenvalues $\lambda_3 = 1$ and $\lambda_4 = -1$ which tell us that

$$\cos^2 \phi_{\text{total}}(C_{\text{XOR}}) = \frac{(1-1)^2 + (0+0)^2}{4} = 0$$

and hence $\phi_{total}(C_{XOR}) = 90°$. Again all we have obtained is a total turning angle which is exactly the angle which bisects the extreme angles of the spectrum of the unitary operator on the complex unit circle.

There is another way to see this limitation on the total turning angle $\phi_{tot}(A)$, and let us note it here. For unitary matrices $A = U$ we have equivalently

$$\cos \phi_{total}(U) = \min_{\|x\|=1} \frac{|\langle Ux, x \rangle|}{\|Ux\| \|x\|} = \min_{\|x\|^2=1} |\langle Ux, x \rangle| = \min_{\lambda \in W(U)} |\lambda|$$

where $W(U)$ is the numerical range of U, e.g., see the book [Gustafson (1997a)]. Because for unitary operators U we know $W(U)$ is exactly the closed convex hull of the spectrum $\sigma(U)$, it follows that its total cosine $|\mu_1(U)|$ is the distance from the origin in the complex plane to the nearest point of $W(U)$. For the example A above the numerical range $W(A)$ is exactly the chord between the two eigenvalues on the unit circle. Thus the total antieigenvectors are just those vectors whose numerical range values $(Ux_\pm, x_\pm)$ attain the midpoint of that chord. When U has eigenvalues 180° apart on the unit circle, as C_{XOR} does, $|\mu_1(U)|$ will always be 0.

To try to gain some intuition for better operator trigonometry for unitary quantum computing operators, let us consider one more here. Let us consider the rotational portions of universal two-bit quantum gate operators such as

$$U = \begin{bmatrix} i \cos \theta & \sin \theta \\ \sin \theta & i \cos \theta \end{bmatrix}.$$

Instead of only looking at pure (real) state vectors $|$off$\rangle$ and $|$on$\rangle$, let us consider U applied to arbitrary (mixed, complex) state vectors $x = (x_1, x_2)$. Then one calculates

$$\langle Ux, x \rangle = 2 \operatorname{Re} (\bar{x}_1, x_2) + i(|x_1|^2 \cos \theta + |x_2|^2 \sin \theta).$$

U has eigenvalues $\lambda_1 = \sin \theta + i \cos \theta$ and $\lambda_2 = -\sin \theta + i \cos \theta$ and so the total cosine is attained by those x which make the imaginary part of this expression vanish.

Thus we admit that the operator trigonometry applied profitably to quantum computing operators needs more work and thought. The angle $\phi_{total}(U)$ seems too crude. The angles $\phi_{re}(U)$ and $\phi_{imag}(U)$ also seem insufficient.

7.4   Penrose Twistors

Without further ado, we consider the twistor key matrix relation [Penrose (2005)]

$$\begin{bmatrix} Z^0 \\ Z^1 \end{bmatrix} = \frac{i}{\sqrt{2}} \begin{bmatrix} t+z & x+iy \\ x-iy & t-z \end{bmatrix} \begin{bmatrix} Z^2 \\ Z^3 \end{bmatrix}.$$

Here t, x, y, z are standard Minkowski coordinates for a point R in four-dimensional real spacetime M, and the twistor space T is the four-dimensional complex vector space with coordinates Z^0, Z^1, Z^2, Z^3 for a point Z there. The relation represents a twistor which is incident with the spacetime point R. The speed of light has been taken as $c = 1$, so we may think in terms of the usual 45° picture of the standard light cone. Let us call the operator $A = i2^{-1/2}H$, where H is the matrix therein. Clearly H is Hermitian, and all spectral properties of A follow from those of H. That is, A and H have the same eigenvectors, the eigenvalues of A are just $i2^{-1/2}$ times those of H. A is a pure imaginary normal operator, $A^* = -A$, $A^*A = -A^2$.

Other twistor operators also depend principally on Hermitian matrices resembling H. I mention a few of these at the end of this section. I also advise the reader to consult the remarkable book of Penrose (2005) for the larger role that twistors may play within fundamental physics. Twistors may be viewed as light rays, as higher-dimensional spinors, as spinning massless particles, within sheaf cohomology, within a nonlinear graviton theory. Penrose's book (2005) is worth buying just for his sketches of the geometry (as he sees it) of fundamental physics.

Here in Fig. 7.2 I have drawn a simple schematic of the twistor geometry which will be adequate for our following discussion. In Part (a) we introduce the notation ψ for the spacetime angle of any point R in real four-space. Part (b), relating the spacetime angle ψ to our operator-trigonometric angle $\phi(H)$, will be explained below.

We wish H's operator trigonometry. For this, we first need H's eigenvalues and eigenvectors. For any point R in spacetime, H has eigenvalues

$$\lambda_1 = t - r, \quad \lambda_2 = t + r$$

where $r = (x^2 + y^2 + z^2)^{1/2}$. For the operator trigonometry, which is more complete for Hermitian positive definite operators, we prefer to consider first and principally points R inside the light cone. There are two cases to be

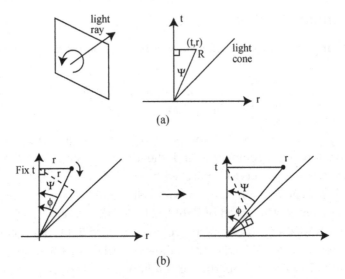

(a)

(b)

Fig. 7.2. Twistor and operator angles.
(a) Twistor spacetime angle ψ for spacetime point R in four dimensions.
(b) Twistor operator angle $\phi(H)$ always exceeds spacetime angle $\psi(R)$.

considered: (a) $r \neq z$, (b) $r = z$. The latter is actually easier but the former is more interesting and general, so we consider case (a) first. Let $EV1$ and $EV2$ describe eigenvectors corresponding to the smaller and larger eigenvalues respectively, where $0 < \lambda_1 < \lambda_2$. These eigenvectors are found to be (up to arbitrary multiplicative constants, of course)

$$EV1 = \begin{bmatrix} (x+iy)/(r+z) \\ -1 \end{bmatrix} = \frac{1}{\sqrt{2r}} \begin{bmatrix} (x+iy)/\sqrt{r+z} \\ -\sqrt{r+z} \end{bmatrix},$$

$$EV2 = \begin{bmatrix} (x+iy)/(r-z) \\ 1 \end{bmatrix} = \frac{1}{\sqrt{2r}} \begin{bmatrix} (x+iy)/\sqrt{r-z} \\ \sqrt{r-z} \end{bmatrix},$$

where I have given two versions of the eigenvectors, the first as they come naturally from their computation, the second when the first are normalized to $\|EV\| = 1$. From the latter we may now calculate the two antieigen-vectors $x^{\pm}$ from their general expression

$$x^{\pm} = \left(\frac{|\lambda_2|}{|\lambda_1| + |\lambda_2|} \right)^{1/2} x_1 \pm \left(\frac{|\lambda_1|}{|\lambda_1| + |\lambda_2|} \right)^{1/2} x_2,$$

where in the present twistor context, x_1 denotes $EV1$ and x_2 denotes $EV2$ as normalized to their norm-one versions. We find the two twistor antieigen-vectors $x^{\pm}$, that is, the two vector directions most turned by H, to be

$$x^{\pm} = \frac{(1 + r/t)^{1/2}}{\sqrt{2}\sqrt{2r}} \begin{bmatrix} (x + iy)/\sqrt{r+z} \\ -\sqrt{r+z} \end{bmatrix} \pm \frac{(1 - r/t)^{1/2}}{\sqrt{2}\sqrt{2r}} \begin{bmatrix} (x + iy)/\sqrt{r-z} \\ \sqrt{r-z} \end{bmatrix}$$

$$= \frac{1}{2}\left(\frac{t+r}{tr}\right)^{1/2} \begin{bmatrix} (x + iy)/(r+z)^{1/2} \\ -(r+z)^{1/2} \end{bmatrix} \pm \frac{1}{2}(t - r/tr)^{1/2} \begin{bmatrix} (x + iy)/(r-z)^{1/2} \\ (r-z)^{1/2} \end{bmatrix}$$

$$= \frac{1}{2(tr)^{1/2}} \begin{bmatrix} (x + iy)\left[\left(\frac{t+r}{r+z}\right)^{1/2} \pm \left(\frac{t-r}{r-z}\right)^{1/2}\right] \\ (-1)[((t+r)(r+z))^{1/2} \mp ((t-r)(r-z))^{1/2}] \end{bmatrix}$$

where I have given three expressions here, all of norm one. The first is by direct calculation, the second emphasizes the $t + r$ and $t - r$ spacetime cross-weighting of the antieigenvector's eigenvector components, the third writes $x^{\pm}$ as a single vector with first spatial component $x + iy$ weighted by a spacetime factor, and the second component -1 likewise.

The corresponding first antieigenvalue μ_1 could have been obtained, before we found $x^{\pm}$, from the general expression

$$\mu_1 = \cos \phi(H) = \frac{2\sqrt{\lambda_1 \lambda_2}}{\lambda_1 + \lambda_2}$$

where $\phi(H)$ is the twistor Hamiltonian H's maximum turning angle. Thus we have for the twistor Hamiltonian

$$\mu_1 = \left[1 - \left(\frac{r}{t}\right)^2\right]^{1/2}$$

$$= [1 - \tan^2 \psi]^{1/2}$$

where ψ is the angle between the t-axis and the point (x, y, z) inside the light cone. But by the Min-Max Theorem of the operator trigonometry, one always has $\cos^2 \phi(H) + \sin^2 \phi(H) = 1$, where $\sin \phi(H)$ is defined by the operator norm minimum

$$\nu_1 = \sin \phi(H) = \min_{\epsilon > 0} \|\epsilon H - I\|.$$

This minimum is attained at $\epsilon_m = 2/(\lambda_1 + \lambda_2)$. Thus from the Min-Max Theorem we have obtained a result linking the twistor operator trigonometry to the spacetime geometric trigonometry.

Theorem 7.1 (Twistor Turning Angle). *Let $\phi(H)$ denote the twistor Hamiltonian H's operator-theoretic maximal turning angle. Let $\psi(R)$ be the spacetime angle of R described above. Then*

$$\sin \phi(H) = \tan \psi(R).$$

We remark that this interesting relation also can be obtained from the fact that for Hermitian operators we know generally from the operator trigonometry that, in the present 2×2 matrix context,

$$\sin \phi(H) = \frac{\lambda_2 - \lambda_1}{\lambda_2 + \lambda_1} = \frac{(t+r) - (t-r)}{(t+r) + (t-r)} = \frac{r}{t} = \tan \psi(R).$$

We may also note that the minimum-attaining optimal "relaxation value" ϵ_m mentioned above for the operator norm minimization turns out to be the inverse of time:

$$\epsilon_m = \frac{2}{\lambda_1 + \lambda_2} = \frac{1}{t}.$$

Also we may interpret the antieigenvalue μ_1 directly in terms of spacetime. In its numerator, the geometric mean expression is the Minkowski pseudometric

$$\sqrt{\lambda_1 \lambda_2} = \sqrt{t^2 - r^2},$$

and since μ_1 is the geometric mean divided by the arithmetic mean of the eigenvalues of H, we have

$$\mu_1 = \cos \phi(H) = \frac{\text{Minkowski pseudometric}}{\text{time}}.$$

Thus we have established intimate connections between the new twistor operator trigonometry and geometric spacetime.

Next let us consider the above-mentioned case (b) when $r = z$. Then $x = y = 0$, $EV1$ and $EV2$ are just

$$EV1 = \begin{bmatrix} 0 \\ 1 \end{bmatrix}, \qquad EV2 = \begin{bmatrix} 1 \\ 0 \end{bmatrix}.$$

Thus $r = z$ is a much simpler situation. The (normalized) antieigenvector expressions become

$$x^{\pm} = \frac{1}{(\lambda_1 + \lambda_2)^{1/2}} \begin{bmatrix} \pm \lambda_1^{1/2} \\ \lambda_2^{1/2} \end{bmatrix} = \frac{1}{\sqrt{2t}} \begin{bmatrix} \pm(t-z)^{1/2} \\ (t+z)^{1/2} \end{bmatrix}.$$

$$= \frac{1}{\sqrt{2}} \begin{bmatrix} \pm(1 - r/t)^{1/2} \\ (1 + r/t)^{1/2} \end{bmatrix}$$

$$= \frac{1}{\sqrt{2}} \begin{bmatrix} \pm(1 - \tan \psi(R))^{1/2} \\ (1 + \tan \psi(R))^{1/2} \end{bmatrix} = \frac{1}{\sqrt{2}} \begin{bmatrix} \pm(1 - \sin \phi(H))^{1/2} \\ (1 + \sin \phi(H))^{1/2} \end{bmatrix}$$

where I have listed several equivalent expressions. These each show varying links between the twistor antieigenvectors and the spacetime geometry.

My general operator trigonometry is less definitive when the operator under consideration loses positive definiteness. One must decide what one wants to do with the null space, in each application, as λ_1 becomes zero. Further, as λ_1 becomes negative, other issues arise. Often then the physical problem or application lacks interest or changes its nature. Nonetheless, for general invertible operators T, one can obtain a general operator trigonometry by defining [Gustafson (2000d)] the angle $\phi(T)$ to be $\phi(|T|)$ where $T = U|T|$ is in polar form. Then the absolute value operator $|T|$ is Hermitian positive definite and the previous operator trigonometry goes through. Whether or not that fits the application at hand needs to be determined in each instance.

Recognizing the just described general reservations, let us then turn to the twistor Hamiltonian when we are on the light cone: $r = t$. Then the eigenvalues are $0 = \lambda_1 < \lambda_2 = 2r$. For case (a) when $r \neq z$, the eigenvectors are as before. However, H takes $EV1$ directly to the null vector. So there is no way $EV1$ can combine with $EV2$ to produce maximally turned antieigenvectors. This is also evidenced in the antieigenvector expressions, wherein now the second terms vanish. Of course for higher-dimensional Hamiltonians, e.g., 4×4 spinor matrices, one could look at the secondary antieigenvalues and their corresponding secondary antieigenvectors but we do not do so here.

Thus what happens in the twistor spacetime geometry is the following. For fixed t and $r \neq z$, as r approaches the light cone from within, the first antieigenvalue $\cos \phi(H)$ approaches 0 as λ_1 approaches 0, the two antieigenvectors $x^{\pm}$ coalesce to $EV1$, the operator turning angle $\phi(H)$ approaches 90°. The expression $\sin \phi(H) = \tan \psi(R)$ still holds but is interpreted as meaning that the twistor Hamiltonian H is free to turn vectors in angles up to 90°.

In case (b) when $r = z$ and we are on the light cone, the situation is simpler and one may check that as before, the eigenvectors are the natural basis. In fact, this is the case whether one is inside, on, or outside the light cone, in case (b) when $r = z$. Then the Hamiltonian H is already in diagonal form. Again the antieigenvector expression collapses to $EV1$ when we are on the light cone.

Let us now return to the generic situation inside the light cone. The result $\sin \phi(H) = \tan \psi(R)$ of Theorem 7.1 naturally raises the question: "How shall we, if possible, conceptually place the operator turning angle $\phi(H)$ into the spacetime geometry?" One way to do this, geometrically, is to start with the spacetime right triangle of the point $(t, r) : (0, 0), (t, 0), (t, r)$, and swing the side which joins $(t, 0)$ to (t, r) down to a side joining $(t, 0)$ to a point (t_1, r_1) such that the triangle $(0, 0), (t, 0), (t_1, r_1)$ is a right triangle. See Fig. 7.2(b). Now we have constructed the point (t_1, r_1), where $t_1 < t$ and $r_1 < r$, such that the triangle $(0, 0), (t, 0), (t_1, r_1)$ is right, with hypotenuse t, and far side of length r. The origin angle ϕ of this second triangle has $\sin \phi = r/t$, which was the tangent of the original triangle. Thus this angle ϕ equals the operator turning $\phi(H)$ according to Theorem 7.1. One always has the operator angle $\phi(H)$ greater than the spacetime angle $\psi(R)$, and the just-described construction is one way to place them both geometrically within the spacetime geometry. Although this construction does not sufficiently elucidate, in our opinion, the twistor operator turning angle dynamics of $\phi(H)$, nonetheless because all of this twistor trigonometry is new, let us record the above considerations as follows.

Corollary 7.1. *The twistor Hamiltonian H's operator-theoretic maximal turning angle $\phi(H)$ always exceeds the spacetime angle $\psi(R)$. As R approaches the light cone, the ratio $\phi(H)/\psi(R)$ approaches 2.*

We have established a new trigonometry for twistors, namely, its operator-theoretic turning angle trigonometry. In so doing, we have established links between the spacetime geometry and the new twistor operator trigonometry. However, this new twistor trigonometry is just at its beginning, and its physical importance is not yet clear. Here we make just a few preliminary observations toward physical uses.

The basic twistor Hamiltonian H occurs in modified forms throughout twistor theory. For example in the interpretation of twistors as spinning massless particles one writes $Z = (Z^0, Z^1, Z^2, Z^3) = (\omega, \pi)$ as two-spinor-parts and the incidence relation between the twistor Z and the spacetime point R may be written as $\omega = ir\pi$ where $r = H/\sqrt{2}$. The spinor $\pi = (Z^2, Z^3)$ represents four-momentum and the spinor $\omega = (Z^0, Z^1)$ represents a part of the particle's six-momentum. The helicity $s = 1/2 \, \overline{Z} \cdot Z$ is essentially, in our notation here, Re $\langle H\pi, \pi \rangle$. Because in the operator trigonometry, the first antieigenvalue μ_1 is generally defined by

$$\mu_1 = \inf_{\pi \neq 0} \frac{\text{Re} \, \langle H\pi, \pi \rangle}{\|H\pi\| \|\pi\|}$$

one could explore the complex trigonometric nature of twistors more generally.

When considering density matrices for particles of spin $1/2$ two-state quantum systems as represented by the Bloch sphere, see [Penrose (2005)], the general Hermitian matrix of trace unity is

$$H = \frac{1}{2} \begin{bmatrix} 1 + a & b + ic \\ b - ic & 1 - a \end{bmatrix}.$$

Clearly this is our H above with a replacing z and time t taken to be 1. To be a density matrix, H must be nonnegative definite, which is the same situation we asked for, being in or on the light cone. Thus our considerations above all apply to these spin $1/2$ density matrices and provide an operator trigonometry for them. Further aspects of this analogy could be pursued elsewhere. In particular, the Bloch sphere setting is quantum mechanical and hence would permit quantum mechanical interpretations of the twistor operator trigonometry we have founded here.

In the general spinor and twistor theory, one finds other similar twistor or spinor Hamiltonians. For example, Lorentzian spinors

$$\frac{1}{\sqrt{2}} \begin{bmatrix} V^0 + V^3 & V^1 + iV^2 \\ V^1 - iV^2 & V^0 - V^3 \end{bmatrix},$$

Weyl spinors, trace-free Ricci tensors, and the like, would all be susceptible to the same operator trigonometric treatment that we have provided here. As Penrose points out, Dirac's four-component spinor formulation often prevails when in fact often it is better to use van der Waerden's later two-component spinors. On the other hand, we could apply our operator trigonometric turning angle approach to Dirac's four-dimensional setting and other, e.g., six- or eight-dimensional twistor settings as well. One simply replaces λ_2 and $EV2$ respectively by λ_4 and $EV4$ where λ_4 is the largest eigenvalue of the matrix in question. We may then also consider the other critical internal operator turning angles.

A comment. In any linear eigenvalue problem, the eigenvalues are at least tacitly, implicitly, in the eigenvectors, even if not appearing explicitly so. However, in this twistor setting, one does not see in the eigenvectors the time variables at all. In contrast the time variable plays an explicit role in the antieigenvectors. Thus the antieigenvector's twisting ability directly reflects time, and hence, the physical dynamics. Of course this may happen more generally when one has a zero trace condition. But it seems of special interest in the twistor context.

This new twistor trigonometry was reported recently in the papers [Gustafson (2009a,b)].

7.5   Elementary Particles

I recently, see [Gustafson (2002b; 2009a)], started an investigation of my operator trigonometry as it might usefully apply to elementary particle physics. The trigger was the fine article by B. Schwarzschild (1999) describing recent experiments physically demonstrating CP violation.

First let us stay in the CP symmetry situation. In the neutral kaon situation we have

$$K_L^\circ = K_2 = \frac{K^\circ - \bar{K}^\circ}{\sqrt{2}} = \text{longer-lived mass eigenstate,}$$

$$K_S^\circ = K_1 = \frac{K^\circ + \bar{K}^\circ}{\sqrt{2}} = \text{shorter-lived mass eigenstate,}$$

where K° and $\bar{K}^\circ$ are the two mass eigenstates of strangeness.

Theorem 7.2 (CP Eigenstates). *The two CP eigenstates K_L° and K_S° are exactly the two strangeness total antieigenvectors.*

This follows from the extension of my operator trigonometry from Hermitian to normal operators for the total antieigenvectors. Here we have $|\lambda_1| = |\lambda_2| = 1$ for the strangeness mass eigenstates K° and $\bar{K}^\circ$. Thus the weak interaction eigenstates K_L° and K_S° are seen as the antieigenstates of the strong interaction basis eigenstates.

We may notice that the above relations may be written in system form

$$\begin{bmatrix} \cos\theta & \sin\theta \\ -\sin\theta & \cos\theta \end{bmatrix} \begin{bmatrix} \bar{K}_\circ \\ K_\circ \end{bmatrix} = \begin{bmatrix} K_1 \\ K_2 \end{bmatrix} = \begin{bmatrix} K_S^\circ \\ K_L^\circ \end{bmatrix}.$$

There the matrix is a CKM (Cabibbo–Kobayashi–Maskawa) quark mixing matrix, with uniform rotation angle $\theta = 45°$.

This CKM quark mixing matrix has (complex) eigenvectors

$$EV1 = \frac{1}{\sqrt{2}} \begin{bmatrix} 1 \\ i \end{bmatrix}, \quad EV2 = \frac{1}{\sqrt{2}} \begin{bmatrix} 1 \\ -i \end{bmatrix}$$

and antieigenvectors (one real, one complex)

$$x^+ = \begin{bmatrix} 1 \\ 0 \end{bmatrix}, \quad x^- = i \begin{bmatrix} 0 \\ 1 \end{bmatrix}.$$

This follows from the eigenvalues $\lambda_1 = e^{i\theta}$ and $\lambda_2 = e^{-i\theta}$. Note, in relation to my last comment in the preceding section, we expect the complex-valuedness of the eigenvalues to be reflected in the corresponding eigenvectors. In the twistor trigonometry we saw that within the spacetime cone, the antieigenvectors were complex-valued also. However, on the spacetime cone, the twistor eigenvectors and antieigenvectors became real-valued. Here, in the quark mixing elementary particle model, we see that one antieigenvector is real-valued, whereas the other one contains a 90° complex phase angle times a real vector. The real and total antieigenvalues are both $\mu = \cos\theta$, as seen from the computations

$$\frac{|\langle Ux^\pm, x^\pm \rangle|}{\|Ux^\pm\|\|x^\pm\|} = \frac{\mathrm{Re}\,\langle Ux^\pm, x^\pm \rangle}{\|Ux^\pm\|\|x^\pm\|} = \cos\theta.$$

If we now postulate a CP violation with a small mixing parameter $\epsilon > 0$, the system is replaced with the system

$$\frac{1}{\sqrt{2}} \left\{ \begin{bmatrix} 1 & 1 \\ -1 & 1 \end{bmatrix} + \epsilon \begin{bmatrix} 1 & -1 \\ 1 & 1 \end{bmatrix} \right\} \begin{bmatrix} \bar{K}_\circ \\ K_\circ \end{bmatrix} = \begin{bmatrix} K_S^\circ \\ K_L^\circ \end{bmatrix}$$

from which one may conclude:

Theorem 7.3 (CP Antieigenvector-Breaking). *CP violation with mixing parameter $\epsilon > 0$ implies that the kaon mass eigenstates K_L° and K_S° are no longer the strangeness antieigenvectors. Thus CP violation may be seen as antieigenvector-breaking.*

This follows from the quark mixing matrix $M = U + \epsilon U^T$ above having eigenvalues $\lambda_{1,2} = [(1 + \epsilon) \pm i(1 - \epsilon)]/\sqrt{2}$ which lie on a radius $(1 + \epsilon^2)^{1/2}$ just outside the unit circle. However M is still a normal operator, so it has two antieigenvectors and we may interpret the CP violating eigenstates as perturbed strangeness antieigenvectors. For example if one takes the experimental value $\epsilon = 0.0023$, one finds the operator trigonometric turning angle has been reduced from 45° to 44.868°.

Recent experimental results [Schwarzschild (2003)] from the Kamland neutrino detector in Japan indicate that neutrino mass eigenstates considered as coherent superpositions of flavor eigenstates have a best-fit mixing angle of $\theta = 45°$. That corresponds to the CP symmetry kaon situation we treated above. Other neutrino mixing angle measurements could involve CP violation but that is not the point I make here. The point is that on average, the content of Theorem 7.2 is experimentally supported, inasmuch as the mixing angle θ is the angle by which the lighter solar neutrino electron state is rotated to the heavier solar neutrino mass state.

My results here are quite incomplete and at the same time (in my opinion) pushing the envelope of what I should claim physically from my operator trigonometry. Still, I find it intriguing to try to tie these fundamentals of constituent matter to my operator trigonometry.

7.6   Trigonometry of Quantum States

By following the recent book [Bengtsson and Zyczkowski (2008)] which provides a geometry of quantum states, I was able to add to that geometry some of my operator trigonometry. The resulting new trigonometry of quantum states was reported in the very recent papers [Gustafson (2010c; 2011a)]. In this section I shall summarize some of those results, closely following those two papers, to which I refer the reader for more details and bibliography.

Entanglement. Consider the two mixed states

$$
\sigma_1 = \frac{1}{3}\begin{bmatrix} 1 & 0 & 0 & 0 \\ 0 & 1 & 1 & 0 \\ 0 & 1 & 1 & 0 \\ 0 & 0 & 0 & 0 \end{bmatrix}, \quad \sigma_2 = \frac{1}{3}\begin{bmatrix} 1 & 0 & 0 & 0 \\ 0 & 0 & 0 & 0 \\ 0 & 0 & 0 & 0 \\ 0 & 0 & 0 & 2 \end{bmatrix}.
$$

The state σ_1 is entangled and the state σ_2 is separable, even though they are isospectral: both have the eigenvalues 0, 0, 1/3, 2/3. Thus both have the same operator turning angle (I just factored out the nullspace of σ_2). But they have different antieigenvectors. Namely, we find the antieigenvectors

$$x^{\pm}(\sigma_2) = \pm \left(\frac{2}{3}\right)^{1/2} \begin{bmatrix} 1 \\ 0 \\ 0 \\ 0 \end{bmatrix} + \left(\frac{1}{3}\right)^{1/2} \begin{bmatrix} 0 \\ 0 \\ 0 \\ 1 \end{bmatrix},$$

whereas

$$x^{\pm}(\sigma_1) = \pm \left(\frac{2}{3}\right)^{1/2} \begin{bmatrix} 1 \\ 0 \\ 0 \\ 0 \end{bmatrix} + \left(\frac{1}{3}\right)^{1/2} \begin{bmatrix} 0 \\ 1 \\ 1 \\ 0 \end{bmatrix}.$$

Thus antieigenvectors serve to distinguish the entangled state σ_1 from the separable state σ_2. We found a similar behavior for twisters in Sec. 7.4, whose dynamics are more related to their antieigenvectors rather than just to their eigenvectors.

Entropy. Entanglement entropy is taken to be the Von Neumann entropy of the reduced state, and hence is the Shannon entropy of the Schmidt vector

$$E(|\psi\rangle) \equiv S(\rho_A) = S(\vec{\lambda}) = -\sum_{i=1}^{N} \lambda_i \ln \lambda_i.$$

For maximally entangled states the entanglement entropy is $\ln N$ whereas it is 0 for separable states.

Two associated measures of entanglement entropy are the tangle

$$\tau(|\psi\rangle) \equiv 2(1 - \mathrm{tr}(\rho_A^2)) = 2\left(1 - \sum_{i=1}^{N} \lambda_i^2\right)$$

and the concurrence $C = \sqrt{\tau}$. Let us consider the two-qubit pure state $|\psi\rangle$ which has tangle and concurrence

$$\tau = C^2 = 2(1 - \lambda_1^2 - \lambda_2^2) = \frac{4\lambda_1 \lambda_2}{\lambda_1 + \lambda_2}.$$

I used the fact that the matrix $\Gamma\Gamma^*$ has trace $1 = \lambda_1 + \lambda_2$ in manipulating the expression above, and we may do so again to arrive at the concurrence

$$C = \frac{2\sqrt{\lambda_1\lambda_2}}{\sqrt{\lambda_1 + \lambda_2}} = \frac{2\sqrt{\lambda_1\lambda_2}}{\lambda_1 + \lambda_2} = \cos\phi(\Gamma\Gamma^*),$$

now expressed operator-trigonometrically. Also, using my Min-Max Theorem, we have the tangle τ expressed operator-trigonometrically, namely,

$$\tau = \cos^2\phi(\Gamma\Gamma^*) = 1 - \sin^2(\Gamma\Gamma^*).$$

Then the entropy of entanglement $E(|\psi\rangle)$ also becomes trigonometric. First we note that the eigenvalues may be expressed as

$$\lambda_1 = \frac{(1 - \sqrt{1 - C^2})}{2} = \frac{1 - \sin\phi(\Gamma\Gamma^*)}{2},$$

$$\lambda_2 = \frac{1 + \sin\phi(\Gamma\Gamma^*)}{2}.$$

These may be placed into the above Shannon entropy formula to render the entanglement entropy $E(|\psi\rangle)$ trigonometric. Thus all three entropy-related measures of entanglement, E, C, and τ, are operator trigonometric.

Disentanglement and Decoherence. Disentanglement as seen by the reduction of concurrence and decoherence exhibited by decay of off-diagonal elements in the density matrix, were investigated in [Jaeger and Ann (2008)] which we shall follow. In particular, one is interested in whether "entanglement sudden death" may occur in a qubit-qutrit system. This leads to a 6×6 time-evolved density matrix

$$\rho'_{AB}(x, t) = \begin{bmatrix} 1/4 & 0 & 0 & 0 & 0 & x\gamma_A\gamma_B \\ 0 & 1/8 & 0 & 0 & 0 & 0 \\ 0 & 0 & 1/8 & 0 & 0 & 0 \\ 0 & 0 & 0 & 1/8 & 0 & 0 \\ 0 & 0 & 0 & 0 & 1/8 & 0 \\ x\gamma_A\gamma_B & 0 & 0 & 0 & 0 & 1/4 \end{bmatrix}$$

where $0 \overset{\leq}{=} x\gamma_A\gamma_B \overset{\leq}{=} 1/4$ to satisfy the positive semidefiniteness requirement. The noise coupling constants are assumed to be $0 \overset{\leq}{=} \gamma_A, \gamma_B \overset{\leq}{=} 1$. The matrix ρ'_{AB} has a four-fold eigenvalue $1/8$ and two more eigenvalues $(1 \pm 4x\gamma_A\gamma_B)/4$. The partial transpose density matrix with respect to qubit A has two-fold eigenvalues $1/4$ and $1/8$ and two more eigenvalues $(1 \pm 8x\gamma_A\gamma_B)/8$. Thus $0 \leq x \leq 1/4$ guarantees positive semidefiniteness for ρ'_{AB} but the partial transpose density matrix may have a negative eigenvalue when x falls within the subinterval $1/8 \leq x \leq 1/4$. The interpretation is that coupling to the environment as represented by the off-diagonal terms $x\gamma_A\gamma_B$ can lead to "entanglement sudden death" when such a negative eigenvalue in the partial transpose density matrix occurs. Here we do not want to immerse ourselves into such model interpretations, but rather just use them to develop more trigonometry of quantum states.

To that end, we take $\gamma_A = \gamma_B = 1$ in ρ'_{AB} and compute its maximal operator turning angle via

$$\sin\phi(\rho'_{AB}) = \frac{(1/4+x)-1/8}{(1/4+x)+1/8} = \frac{8x+1}{8x+3}.$$

For $x = 1/8$ we see that $\sin\phi(\rho'_{AB}) = 1/2$ and $\phi(\rho'_{AB}) = 30°$. More generally we see that as such finite time disentanglement proceeds, e.g., as x decreases from $1/4$ to 0, the density matrix operator angle decreases from $\phi(\rho'_{AB}) = \sin^{-1}(3/5) = 36.86989765°$ to $\phi(\rho'_{AB}) = \sin^{-1}(1/3) = 19.47122063°$.

Schmidt Angles. Let us consider the Schmidt angle as treated in [Bengtsson and Zyczkowski (2008), p. 336]. Starting with a pure state $|\psi\rangle = \cos\chi|00\rangle + \sin\chi|11\rangle$, the partial trace procedure yields

$$\rho_A = \text{Tr}_B|\psi\rangle\langle\psi| = \begin{bmatrix} \cos^2\chi & 0 \\ 0 & \sin^2\chi \end{bmatrix},$$

where the Schmidt angle $\chi \in [0, \pi/4]$ parametrizes the amount of ignorance of the system, i.e., amount of entanglement. We see immediately that the operator trigonometric angle of the density matrix ρ_A is obtained through

$$\cos\phi(\rho_A) = \frac{2\sqrt{\cos^2\chi\sin^2\chi}}{\cos^2\chi+\sin^2\chi} = \sin 2\chi.$$

Therefore we may conclude: the operator turning angle $\phi(\rho_A) = \pi/2 - 2\chi$ measures entanglement in a way equivalent to that of the Schmidt angle, but in inverse manner. Bigger operator turning angle corresponds to smaller entanglement.

We note in passing that this trigonometric dynamics seems to go in the opposite direction of that of the previous paragraphs. However, the circumstances are different. In the previous paragraphs the amount of entanglement was due to a decoherence coupling with off-diagonal noise. Here we are dealing with partial traces (e.g., starting with Bell states) and then only diagonally-measured entanglement.

Quantum Channels. In the rather complicated paper [Mendl and Wolf (2009)] unital quantum channels are investigated. These are doubly stochastic completely positive transformations mapping the density states into themselves. One important aspect of this theory is that a quantum channel permits perfect environmentally assisted error correction if it is a mixture of unitaries. This property resembles that of the classical Birkhoff Theorem for doubly stochastic matrices.

We take one example therefrom to obtain some new quantum state trigonometry. A minimization problem $\mathrm{tr}[U_S \overline{U}_S^{T_2}]$, given fixed $\mathrm{tr}[U\overline{U}]$, leads to the 9×9 matrix

$$D_S = \begin{bmatrix} \cos\theta & & & & & & & & \\ & \cos\theta & & & & & & & \\ & & \cos\theta & & & & & & \\ & & & \cos\theta & & & & & \\ & & & & \cos\theta & & & & \\ & & & & & \cos\theta & & & \\ & & & & & & \cos\theta & & \\ & & & & & & & \cos\theta & \\ & & & & & & & & 1 \end{bmatrix},$$

which in turn leads to a proposed minimum

$$\frac{1}{d^2}\mathrm{tr}[U_S \overline{U}_S^{T_2}] = \langle G\sigma, \sigma \rangle = \frac{1}{9}\left[-\frac{8}{3}(\cos\theta + 1)^2 + 3 \right]$$

where θ must satisfy

$$\frac{1}{d^2}\mathrm{tr}[U\overline{U}] = \frac{1}{9}[16\cos^2\theta - 7].$$

We note that here θ is a usual submatrix rotation angle coming from a 9×9 matrix

$$D = \bigoplus_{i=1}^{4} \begin{bmatrix} \cos\theta & -\sin\theta \\ \sin\theta & \cos\theta \end{bmatrix} \oplus 1.$$

We may now obtain the operator trigonometry in D_S from

$$\sin\phi(D_S) = \frac{1 - \cos\theta}{1 + \cos\theta} = \tan^2\left(\frac{\theta}{2}\right).$$

The (most turned) antieigenvectors are

$$x_\theta^\pm = \pm \left(\frac{1}{1 + \cos\theta} \right)^{1/2} [x_{1-8}]^T + \left(\frac{\cos\theta}{1 + \cos\theta} \right)^{1/2} [x_9]^T.$$

Here x_{1-8} denotes any norm-one eigenvector from the $\cos\theta$ eight-dimensional eigenspace, and x_9 is $[0,0,0,0,0,0,0,0,1]$.

Quantum Fidelity. Quantum fidelity is a big subject. A usual view, e.g., see [Nielson and Chuang (2000)], implies that quantum fidelity is essentially various versions of the distances between states in the Hilbert–Schmidt inner product $\langle A, B \rangle = \mathrm{tr}(AB^*)$. Thus for example we find that

$$F(\rho, \sigma) = \mathrm{tr}\sqrt{\rho^{1/2}\sigma\rho^{1/2}} = \max |\langle\psi|\phi\rangle|.$$

Our first point to be made here is that in the operator trigonometry we are not interested in the standard angle between vectors, e.g., writing an inner product as $\langle u, v \rangle = \|u\| \|v\| \cos\theta$. In fact the old classical fidelities of the latent semantic index search engines which were so dominant in the (pre-2000) preGoogle era were based essentially on such intervector cosines. And for that reason, we never developed any operator trigonometry for them.

Therefore, let us instead go to the paper [Knill and Laflamme (1997)] where subspace fidelity and entanglement fidelity are compared within a context of error-correcting codes. There an initial quantum state ψ_i undergoes an interaction with the environment to become a reduced density matrix

$$\rho_f = \$(|\psi_i\rangle) = \sum_a A_a \rho_i A_a^*,$$

where $\$$ represents a superoperator associated with the environmental interaction, and the $A_a = \langle \mu_a | U | e \rangle$ couple to an orthonormal basis μ_a of the environment with the environment's initial state e through the evolution operator U of the whole system. Fidelity is then taken to be the "overlap" between the final state ρ_f of a system ρ and the original state ψ_i. This "overlap" viewpoint is the one we were "not interested in" in our comment above. However now we have operator turnings. One arrives at the minimum fidelity

$$F_{\min} = \min_{|\psi\rangle} \langle\psi|\rho_f|\psi\rangle = \min \langle \$\psi, \psi \rangle,$$

and the best quantum code is taken to be that which maximizes $F_{\min}$. One takes the Kraus operators $\{A_a\}$ above to include both environmental interaction and error-correction recovery. To get to my operator trigonometry one ideally wants $F_{\min}$ of the form

$$F_{\min} = \min_{|\psi\rangle} \frac{|\$\psi, \psi\rangle|}{\|\$\psi\|\|\psi\|}$$

so that one can investigate the trigonometric properties of the environmental interaction superoperator $\$$. That situation can be obtained if $\$$ is implemented with Hermitian or unitary operators. Then the criterion for best quantum (error-correcting) code, namely, to maximize $F_{\min}$, now trigonometrically means to maximize the first antieigenvalue of the $\$$ operator. Said another way, it is to minimize the turning angle of $\$$.

Going further, one lets C denote the error correcting code, A_a the polluting operators, and R_r the recovery operators. Then the fidelity of the error-correcting code C is

$$F(C, R, A) = \min_{|\psi\rangle \in C} F(|\psi\rangle, RA) = \min_{|\psi\rangle \in \text{Codes}} \sum_{r,a} |\langle \psi | R_r A_a | \psi \rangle|^2.$$

For this, the operator trigonometry of the operators $R_r A_a$ again apply. We see that, ideally, for perfect error correction, one wants to drive all of these interior trigonometric operator angles to zero.

Orbit Stratification. Here we follow [Bengtsson and Zyczkowski (2008)] and the earlier fundamental paper [Kus and Zyczkowski (2001)]. There are many details we cannot develop here. To illustrate the "Weyl Chamber" geometry, we may take an example. Consider that one has a pure state W represented by a 4×4 matrix $|w\rangle\langle w|$ which leads to a 6×6 Gram matrix

$$C = \left[\frac{1}{2} \text{Tr}(W_m W_n) \right] = \begin{bmatrix} A & B \\ B^T & D \end{bmatrix}.$$

Then one wishes to look for C's orbital invariants under the transformations by the orthogonal group $C' = OCO^T$ to be regarded as various measures of entanglement. The rank of C is the dimension D_ℓ of the orbit. For a pure-state composite 2×2 quantum system one arrives at a six-dimensional submanifold of the 15-dimensional manifold of all density matrices on four-dimensional Hilbert space.

Letting $w = [v, x, y, z]^T$ induce W and then C, we find that C has the six eigenvalues

$$\lambda_1 = 0, \quad \lambda_2 = 8|\omega|^2, \quad \lambda_3 = \lambda_4 = 1 + 2|\omega|, \quad \lambda_5 = \lambda_6 = 1 - 2|\omega|,$$

where $\omega = vz - xy$. The concurrence in this situation is $c = 2|\omega|$. For further analysis of orbit stratification, one maps w to the matrix

$$X(w) = \begin{bmatrix} v & y \\ x & z \end{bmatrix}$$

and W to the representation

$$W_\theta = w_\theta w_\theta^* = \begin{bmatrix} \cos\theta/2 \\ 0 \\ 0 \\ \sin\theta/2 \end{bmatrix} \begin{bmatrix} \cos\dfrac{\theta}{2}, 0, 0, \sin\dfrac{\theta}{2} \end{bmatrix}.$$

SVD is used for the latter. One finally arrives at the 2×2 matrix

$$X' = \begin{bmatrix} p & 0 \\ 0 & q \end{bmatrix} = \begin{bmatrix} \cos\theta & 0 \\ 0 & \sin\theta \end{bmatrix}$$

where $\sin\theta = 2\omega = c$ represents the concurrence.

Let us now put some operator trigonometry into the above (very) geometrical picture. We find, using the trace-one property, that

$$\cos\phi((X')^2) = \frac{2pq}{p^2 + q^2} = 2\cos\theta\sin\theta = \sin 2\theta.$$

A resemblance to our results for Schmidt angle in the above is immediate, but here θ is the concurrence angle. Directing the reader now to the famous Weyl chamber stratification pictures, see [Bengtsson and Zyczkowski (2008)], we may conclude the following for $N = 4$. The three-dimensional submanifold of maximally entangled states corresponds to $\omega = 1/2$ and hence to concurrence $c = \sin\theta = 1$. Trigonometrically, we see that means $\cos\phi((X')^2) = 0$ which implies a 90° turning angle. Then for ω decreasing from $1/2$ to 0, in the interval $1/2 > \omega > 0$ the X' angle θ drops from 90° to 0 but our operator trigonometric $\cos\phi((X')^2)$ increases to 1 at $\theta = 45°$ and then drops to 0 as ω approaches zero. Thus the trigonometric operator turning angle $\phi((X')^2)$ first decreases its 90° turning to no-turning and then increases toward a 90° turning again. Geometrically this corresponds to a five-dimensional orbit movement through a five-dimensional manifold of nonmaximally entangled states. At $\omega = 0$ one has a four-dimensional manifold of separable states. Those have a trigonometric turning angle $\phi((X')^2)$ of $90°$, as we would expect for rank-deficient pure separable states.

There is certainly a lot more new quantum state trigonometry to be found in the quantum state orbit stratification theory. For example, one could seek all critical higher antieigenvalue turning angles in the 6×6 matrix C. But such is too complicated to pursue here.

Commentary

The fundamental and inescapable role of chance in our lives as well as in quantum mechanics is a recurring theme in my autobiography [Gustafson (2011e)]. But that need not be elaborated here. Quantum mechanics will continue to require our probabilistic mathematical models in its phenomena until we are able to really see what an electron is, what photons are, climb onto a spinning proton, by some exquisite as yet unperformed brilliant experiments.

Nonetheless, my contributions to the Bell inequality theory and my new quantum spin identities in Secs. 7.1 and 7.2 do replace a number of philosophically-based claims about nonlocality and hidden variables by a beautiful new mathematical vector trigonometry that serves to clarify the objective meanings of the situation.

My preliminary dabbling into quantum computing via its finite rank unitary operator description in Sec. 7.3 no doubt has revealed to all readers the inadequacy of my operator trigonometry beyond Hermitian operators. Stated another way, to date my intuition has not yielded any breakthroughs as to how I should actually "see" complex-valued angles. I still see angles as angles. I do not like to mess them up with phase. But I regard this situation as just a limitation caused by my own stubborn refusal to go beyond what I can see in my mind's eye. I am sure others will soon improve my operator angle theory by just treating it within abstract modern differential geometry and not worrying about its visualizations.

A well-known complaint about Penrose's twistor theory is that it does not yet get mass properly into it. A hoped-for outcome would be the explaining of wave function collapse as somehow caused by gravity or some related effect. My own intuition does not travel along those roads. Instead I think we need a more detailed understanding of the real physics that occurs within a quantum measurement. It is fine to send a "photon" across two meters of Earth's atmosphere and have a massive avalanche diode say: "Hey, you got an event."

An analogy would be the incredible and expensive effort over many years by the United States government to go beyond all the statistical findings undeniably linking tobacco smoking to cancer, to actually find and prove *the chemical causative mechanisms*. That was a great albeit expensive scientific and human achievement. And then the United States government won the legal case hands down, and could start to collect billions of dollars of damages from the tobacco companies.

My very limited operator trigonometry for elementary particles in Sec. 7.5 could be greatly improved by extending my operator trigonometry into the group theory of the Standard Model. This should be done but I doubt that I have enough time to succeed. So I am just claiming some tantalizing glimpses as seen by a member of the audience, so to speak.

By contrast, much more of my operator trigonometry for quantum states as treated in Sec. 7.6 should be expected in the future, because those density matrices are tailor-made for my theory. However, I am not sure at this writing the right questions to ask, to open doors through which one may probe deeper into that quantum geometry.

7.7   Exercises

1. Bell's original inequality

$$|P(a,b) - P(a,c)| \overset{\leq}{=} 1 + P(b,c)$$

actually is a special case of the following rather general inequality which can be useful in a number of probabilistic situations. Let a, b, c be any real numbers in the interval $[-1, 1]$. Then

$$ab - bc + ac \overset{\leq}{=} 1.$$

Prove this.

2. Verify my Wigner inequality–equality in Sec. 7.2. What is the most simple trigonometric inequality–equality you can think of? Should my idea become a new theme throughout mathematics?

3. Can you feel, intuitively, another explanation of why the operator trigonometry in its present state does not yield much profit when treating unitary matrices?

4. In the $r = z$ twistor situation, the antieigenvectors turned out to be

$$x_\pm = \frac{1}{\sqrt{2}} \begin{bmatrix} \pm(1 - \sin \phi(H))^{1/2} \\ (1 + \sin \phi(H))^{1/2} \end{bmatrix}.$$

Show that this gives us a general new operator-theoretic identity valid for positive Hermitian matrices.

$$\left[\frac{1 - \sin \phi(H)}{1 + \sin \phi(H)} \right]^{1/2} = \frac{\cos \phi(H)}{1 + \sin \phi(H)} = \frac{1}{\kappa(H^{1/2})}.$$

This identity complements an operator–theoretic identity we have previously [Gustafson (1997b)] found useful,

$$\left[\frac{1 - \cos \phi(H)}{1 + \cos \phi(H)} \right]^{1/2} = \frac{\sin \phi(H)}{1 + \cos \phi(H)} = \sin \phi(H^{1/2}).$$

5. Show that with the experimental value $\epsilon = 0.00023$ for the mixing parameter, the quark mixing matrix of Sec. 7.5 has operator angle 44.868°.

6. Consider the useful Bloch sphere density state representation (see [Gustafson (2010c)], and citations therein)

$$D(r, \theta, \phi) = \frac{1}{2} \begin{bmatrix} 1 + r \cos \theta & r \sin \theta e^{-i\phi} \\ r \sin \theta e^{i\phi} & 1 - r \cos \theta \end{bmatrix}.$$

Here $0 \leq r \leq 1, 0 \leq \theta \leq \pi, 0 \leq \phi < 2\pi$, and a pure state qubit

$$|\psi\rangle = \cos \frac{\theta}{2} e^{-i\phi/2} |0\rangle + \sin \frac{\theta}{2} e^{i\theta/2} |1\rangle$$

is represented by the point

$$u(1, \theta, \phi) = (\sin \theta \cos \phi, \sin \theta \sin \phi, \cos \theta)$$

on the surface of the Bloch sphere. An arbitrary spin density state in this representation corresponds to a point $u(r, \theta, \phi)$ in the Bloch sphere which is the convex combination

$$u(r, \theta, \phi) = r u(1, \theta, \phi) + (1 - r) u(0, \theta, \phi)$$

of a pure state and the sphere center density state. Develop some of the operator trigonometry for D.

Chapter 8

Financial Instruments

Perspective

Long ago I obtained combined Engineering-Business Bachelor of Science degrees from the University of Colorado at Boulder. My Engineering majors were Physics and Applied Mathematics and my Business major was Finance. Shortly after, I was compelled to serve four years military service. Unusual circumstances found me spending much of that time in pioneering computing design for military intelligence. I have given an unclassified account of some of that work in [Gustafson (1999h)]. Not told until now was my critical role in the Cold War of writing the software for the world's first electronic intelligence satellite in 1960. That and other interesting aspects of my life may be found in my autobiography [Gustafson (2011e)]. While I was engaged in that military obligation and early computer science, in 1965 I obtained my Ph.D. in Mathematics from the University of Maryland. That was followed by two postdoctoral years in Europe which led me into the life of an academic professor of mathematics.

This engineering, business, physics, mathematics, and computing background augmented by several years working in military electronic intelligence, and even by one year of Law School, has provided me with a much wider background than most who work in Finance. But I never thought I would become involved with financial derivatives.

I want to continue this brief history here for a specific reason. First, what I will present in this chapter, the application of my operator trigonometry to financial instruments, which albeit my very narrow expertise, notably, that of my operator trigonometry, addresses the financial modeling industry which is vast, and I do not think I need to say, is of great importance to human society. Therefore, I want to advertise an opinion to the effect that we need to require more breadth, over depth, which

I mischievously at times call narrowness, in our mathematics majors at universities.

A first step would be to bring back the requirement of a strong minor in a field distinct from mathematics. For example, there are still a significant number of physics majors who are also getting a mathematics minor and in many cases a mathematics major too. That is even better. To cut short further rhetoric here, I want our mathematics majors to be more widely trained and therefore more widely competent and consequently more valuable, like me! In that way they will be more able to later partner with engineers, physicists, economists, biologists, educators, and experts in other fields of human endeavor.

To continue, during the reinvigorization of the science of neural networks, I helped obtain a $22 million NSF Optical Computing Systems Engineering Research Center Grant for the University of Colorado in 1988. A particular goal was to see if the high parallelism of optics could be married to the high complexity of neural networks to produce a new sixth generation computing paradigm, one which would be more capable of learning and even thinking than are the current fifth generation digital computers. My work with various state-of-the-art neural network algorithms led to an essentially unrelated collaboration with ABB Corporate Research in Baden, Switzerland, whose interest in neural networks was to investigate their potential for modeling financial markets and financial traders. A few million Swiss Francs were available for experimentation by the small team of physicists among whom I was attached as a consultant. We had good months and bad months for a few years. My own feeling became that of viewing weighted neural networks as nonlinear regression algorithms which were just not elastic enough or not truly stochastic enough to capture the intrinsic randomness of the markets. They were also not adequate to capture the varying human thought patterns among human traders.

Our work at ABB surprisingly led us toward psychology: the human factor. We published some interesting papers comparing the performance of neural networks, some of the artificial intelligence community's best learning algorithms, and humans, when faced with certain decision problems. Our papers [Bernasconi and Gustafson (1992; 1994; 1998)] are quite unique in determining how humans reason, simplify, and generalize. One key finding was that humans absolutely must establish a *context*, even an imagined one, before they can reason and make decisions. Thus the ability of a financial trader to make quick decisions depends on the scenarios he can quickly bring to mind in order to find a context for the situation at hand. In particular, our Backprediction criterion of validity of generalizations seems better than any other and gives valuable insights as to why humans do better than any of the black-box concept learning algorithms.

In 1994 I spent one month of a sabbatical semester at ABB Corporate Research commissioned to take a look generally at state-of-the-art portfolio analysis, and in particular, in the just publicly released risk metrics system of J. P. Morgan. I was astonished to see for the first time the famous Black–Scholes partial differential equation, which had come into being about twenty years earlier. In 1998 the Finance Division of the University of Colorado Business School sent me several of their Ph.D. students to enable my teaching the Black–Scholes partial differential equation for pricing options and futures. I also eventually produced two financial Ph.D. students in the Mathematics Department from this effort.

The famous Black–Scholes PDE is

$$\frac{\partial V}{\partial t} + \frac{1}{2}\sigma^2 S^2 \frac{\partial^2 V}{\partial S^2} + rS\frac{\partial V}{\partial S} - rV = 0$$

where typically $V = V(S, t)$ is a call or a put or more complicated option to buy or sell a stock or other asset at a predetermined time T in the future. When I teach Black–Scholes I always quickly present the call and put solution pictures shown here in Fig. 8.1. Those quickly illustrate that the basic option pricing problem in both cases is that of coming back from the payoff values $S - E$ or $E - S$ at the time of expiry $t = T$ to find what you should pay for the option now at $t = 0$. Thus Black–Scholes PDE problems are the motivating example for Backward Stochastic Differential Equations (BSDEs for short). I give a short derivation of them using Ito calculus in my book [Gustafson (1999f)]. I also show there how you can convert them to the heat equation and using its Green function go further to reduce everything to the normal probability distribution. Thus already in the 1970s the Chicago commodities exchange trades could calculate option prices just from the normal probability curve on their hand calculators.

CFD methods (Computational Fluid Dynamics) can be used to model option derivatives. An example is the paper of Zvan, Forsyth, and Vetzel

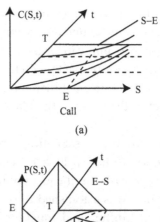

(a)

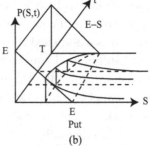

(b)

Fig. 8.1. Call and put options.
(a) European call. T is the maturity date. You solve backward to price at time $t = 0$.
(b) American put. The curve on the left is the exercise boundary. You may exercise your option at any time.

(1998), where a square-root mean-reverting stochastic volatility model

$$dS = \mu S dt + \sqrt{v} S dz_1,$$

$$dv = \kappa(\theta - v)dt + \sigma\sqrt{v}dz_2,$$

leads to a rather complicated valuation PDE of Black–Scholes type which is then solved numerically. That equation may be written conveniently in a two-vector convection–diffusion form as

$$\frac{\partial U}{\partial t} - \nabla \cdot D \cdot \nabla U + V \cdot \nabla U + rU = q.$$

The q is an added penalty term to be used to distinguish whether you are in the PDE region or in the exercise region. There are many powerful CFD numerical methods to discretize and solve such equations ... provided you can come up with appropriate boundary conditions. In Zvan *et al.* a Galerkin finite element basis is introduced and the resulting algebraic system is solved by Newton iteration. Therein a nonsymmetric Jacobian matrix problem is resolved by an ILU preconditioned BICGSTAB linear solver.

What are the boundary conditions to be used? Resort to taking the equation onto the boundary is employed. Instantaneous volatility $\sqrt{v} \to 0$ is modeled by dropping corresponding terms in the equation. Likewise $\sqrt{v} \to \infty$. Likewise treated is the stock price going to zero or infinity. Application is made to American and European chooser options with double knockout barriers. Rather esoteric optimal early exercise regions in the $S - \sigma$ plane are found.

My first mathematical finance Ph.D. student J. Davenport and I worked along such general lines. In particular we showed how the onset of an arbitrage opportunity is related to the violation of the high-contact condition at an exercise boundary when market drift exceeds the riskless rate near a dividend payout. As a result, an exercise boundary can actually turn back on itself. The Geske–Roll–Whaley treatment of dividends was thus seen to possess some fallibilities.

This was all rather high-level analytical and computational partial differential equations applied to the pricing theory of financial options and futures. But the world of financial engineering moves very fast. Vanilla Black–Scholes PDEs were more or less superceded in the 1980s by martingale models and more important parameter considerations such as the empirical volatility smile. Very recently I have been looking into the world of high frequency trading algorithms. Jumping into the bid–ask gap near the bid price and out in a matter of milliseconds near the ask price can make you $millions, even $billions if you have high enough volume. In 2009, 73% of all USA equity trading was performed via HFT trading algorithms.

My second mathematical finance Ph.D. student T. Seguin came to me in 2004 from a financial and banking background. He survived, even did well in my PDE course, then proposed that we see if there was any way we could price venture capital. That market is notoriously incomplete and his question was audacious, to say the least. By chance my daughter knew from her Berkeley law school days a venture capitalist here in Colorado who generously supplied us with all venture capital market data (mostly it was just statistics) which was available to him. Although we backed off from the venture capital pricing goal, nonetheless we learned a lot about the risk measures for incomplete markets that have been introduced during the last ten years, and finished a Ph.D. with an in-depth study of so-called coherent risk measures. Typically these satisfy only the four axioms:

$$\text{Subadditivity} : \rho(X + Y) \le \rho(X) + \rho(Y)$$

$$\text{Homogeneity} : \rho(\lambda X) = \lambda \rho(X)$$

$$\text{Monotonicity} : X \le Y \Rightarrow \rho(Y) \le \rho(X)$$

$$\text{Translation} : \rho(Y + m) = \rho(Y) - m.$$

To fix ideas, think of X and Y as two portfolios of investments. Then, roughly, subadditivity says that larger diversification reduces risk, homogeneity says that increasing the amount of a fixed portfolio increases risk, monotonicity says that a better portfolio will carry less risk, and translation says that if you add riskless assets to your portfolio, you decrease risk. We looked closely into the use of penalty functions to better understand the nature of downside risk.

A good way to view a coherent measure of risk ρ is as follows. You compute the expected loss $E_Q(-X)$ of a position X for each considered measure Q, and then take the worst result. Thus $\rho(X) = \sup E_Q(-X)$ over a hopefully adequate set of measures $[Q]$. Penalty functions α are then introduced onto that set of measures and your risk measure becomes $\rho(X) = \sup[E_Q(-X) - \alpha(Q)]$. Generally the penalty functions are taken to be convex and all of this kind of mathematical risk modeling is a rather overconfident attitude that you can know the set of measures which will describe all events that might impact your portfolio of investments. This is important because the design of your penalty functions depends on that class of measures. In particular, it seems to me that the construction of these penalty functions depends too much on an assumption of translation invariance.

However, neither of my mathematics Ph.D. students who did their Ph.D. dissertation work under my supervision in financial mathematics, nor I, have had time to extract from their investigations papers for publication. Both have moved on to other pursuits. Also, the financial world moves too quickly. Probably CFD, e.g., PDE, methods for pricing options, are too slow now. Monte Carlo, instead, for treating larger portfolios, has come to the fore. I have not had time to follow my own instincts about fallibilities in the risk measure formulations.

I have written, on the other hand, upon invitation from the Shanghai Finance University after I gave a lecture there in June 2010, an account

[Gustafson (2010f)] of my experiences and insights in mathematical finance during the last twenty years. In Sec. 8.1 below I will extract from that material to try to give you some flavor (very brief, and to my tastes) of this exciting subject.

Then quite independent of all of the above-mentioned experiences in mathematical finance comes the main theme of this chapter: the application of my operator trigonometry to certain financial instruments. These models were selected simply because I had some familiarity with them and felt that they might possess some hitherto interesting operator trigonometric features. Specifically, Sec. 8.2 considers a quantos model in which a stock price and a currency exchange rate follow correlated stochastic processes. Section 8.3 treats spread options for assets modeled by Black–Scholes or trinomial tree pricing and hedging algorithms. Section 8.4 involves a Markowitz–Vasicek portfolio interest rate model. Section 8.5 considers a Black–Scholes model with random volatility process. Section 8.6 looks at an important recent paper from the emerging field of risk measures for incomplete markets. I have taken most of this from the recent paper [Gustafson (2010a)].

8.1   Some Remarks on Mathematical Finance

The mathematization of modern finance was inevitable. One starting point was the economic theory of choice under uncertainty as treated by Von Neumann and Morgenstern (1944). In a study [Gustafson (1994d)], I found at least ten Nobel prizes awarded in Economics (Allais, Arrow, Friedman, Harsanyi, Kantorovich, Markowitz, Nash, Samuelson, Selton, Simon) with substantial mathematical content.

An early and forgotten work by Batchelier, a student of Poincaré in France, modeling financial market prices as Brownian motions, was rediscovered by Savage and Samuelson as a precursor to the final 1972 formulation of the Black–Scholes partial differential equations to price options. Good classical treatments of that subject may be found in [Merton (1990); Wilmott (1998)]. A very useful highly mathematical recent treatment of the seemingly ever-present arbitrage in markets is given in [Delbaen and Schachermayer (2006)]. An excellent treatment from the stochastic processes point of view is [Föllmer and Schied (2004)].

As I started teaching Black–Scholes theory in the late 1990s, as a mathematician I developed an increasing unease about a fundamental assumption that is blithely taken in all the treatments I know. As the reader knows, the Black–Scholes theory is so important in finance as to be covered in many, many books, papers, and so on. However for my purposes here I would like to refer the reader specifically to a nice paper [Andreason, Jensen, and Poulsen (1998)]. This paper presents eight different derivations, from differing financial and economic viewpoints, all of which arrive at formulations equivalent to the Black–Scholes PDE.

I was struck with the fact that in virtually all of the literature, the mathematics treats the stock price S and the time t as two independent variables over which you plot the option valuation $V(S, t)$. This is Cartesian thinking. This is how I drew the valuations in Fig. 8.1. Yet everyone knows that stock price S and time t are not really independent in financial markets. However, Black–Scholes theory works because it is derived from the generally valid assumption that the stock price changes ΔS are random. What bothered me was the underlying issue of whether the stochastic independence of causally random price change increments was consistent with the Cartesian independence of algebraically and functionally completely independent variables S and t.

Accordingly I assumed not and wrote a paper entitled *A 9th Derivation of the Black–Scholes equation* [Gustafson (2000f)] and presented it at a conference on Computational Methods in Decision Making and Finance held in Neuchatel, Switzerland, in August 2000. This was a rather detailed and quite mathematical paper of 28 single-spaced pages. Although I would be the first to admit that it was not complete in all respects, nonetheless I looked at all eight valuation methods put forth in [Andreason *et al.* (1998)] and then compared my own rederivation of the Black–Scholes PDE in which I allowed (this was new) some Cartesian dependence. The basic idea I developed was to do the Ito calculus looking at the total change dV/dt rather than $\partial V/\partial t$. My ansatz was that in practice, one needs *some* time to pass before $S(t)$ can adjust itself.

But the referees did not understand it and therefore did not like it and rejected it for publication. Remember that the original Black–Scholes paper was twice rejected for publication! To assure priority for my idea, I presented very briefly my main ideas in my chapter [Gustafson

(2002d)]. I concluded that the very notion of hedging (instantaneously) is a (convenient) myth. I speculated that this myth might be obviated in practice by the rapidity of updated re-pricing. Now I wonder if I was prescient in anticipating the recent practices of flash and naked trading wherein whether a computer-driven high-frequency trade will win or lose depends on being a few hundred microseconds ahead of the other fellow. Such practices, variations on front running but apparently legal, are reputed to have made Goldman–Sachs a large part of their record multibillion dollar profits this last year. Whether that is related to my idea or whether not, are there other interesting variations on my idea?

Next, I turn to financial crashes, and their causes. From a physicist's point of view, the validity of a PDE obtains from the presence of the physical conditions under which it was derived. The validity of the Black–Scholes PDE rests upon an assumption of no-arbitrage, for which the roughly equivalent but more descriptive terminology "no free lunch" is sometimes used. The micro-version of this is that buyers and sellers are so active that whenever they place a bet in a hope of a free profit, that bet gets evened out by opposing bets. Thus all would-be arbitragers cancel each other out and a correct price is produced.

What really makes the science of financial derivatives so fascinating is its coupling to humans using the resulting financial-engineered products, and more to the point, to human nature. For example, it is not so widely known that a major reason for the Crash of 1987 was an aspect of model risk. In that instance, too many of the quants were using essentially the same Black–Scholes equations. So the needed balance of opposing bets was lost. An excellent treatment of this thesis is developed in [MacKenzie and Millo (2003)]. I would like to add here two of my own observations. These emphasize the human nature parts of model risk.

The development of the Chicago Board Options Exchange (CBOE) and the advent of Black–Scholes theory in 1972 essentially coincided and thereafter prospered hand-in-hand. From 1976 until 1986 the Black–Scholes–Merton model was a close-fit to observed market option prices. Then in 1986 the CBOE put in a system called Autoquote which would take existing option prices from the market and generate synthetic stock prices to fill in the gaps in an incomplete market. Thus traders would have a wider range of option prices from which to choose. This created a circular situation in which option prices drive synthetic option prices which in turn drive option prices. But the Black–Scholes equations need to assume that the S in them is known to all and is not dependent upon the option values V. This observation is a variation on my thesis in [Gustafson (2000f; 2002d)]. Not only may the independent variables S and t not really be independent, but even worse, we do not want them to depend on the dependent variable V! So the model drove the market.

Why would the quants or traders permit this to happen? You must make as much money for your employer as your colleague does for his. So you are driven to use the same, latest, financial instrument technology.

My second observation is that in game theory terms, everyone is a free-rider on the way up. But when the crash comes, each human individual bails to cut his losses. Then not only is the delicate buyer–seller balance from which Black–Scholes equation is derived lost, but the market goes into a free-fall. I call this a *suddenly incomplete market*, and it is a possibility that is always hiddenly present. After the October 19, 1987 free-fall of over 20% of market values, the Cox–Ross–Rubenstein improved version beyond the Black–Scholes model came into style to allow the trader some flexibility. It also has built into it a probability of a crash. But no model can survive when all the buyers rush to the exits.

By chance in June 2007, on my way to a quantum physics meeting in Sweden, I stopped into a remarkable yet small three-day retirement conference in Berlin for Professor Hans Föllmer — a noted expert on stochastic finance. There I learned from some of the most highly regarded academic mathematical finance professors that they could not accurately price CDOs (collateralized debt obligations). To my surprise, I also learned that nonetheless the banking community was using those financial instruments extensively, and that very large sums of money — in the billions — were in play. About two months later, on August 10, 2007, the Dow Jones Average fell 387 points and the Credit Panic of 2007 was underway. CDOs and similar financial instruments, the CDSs (credit default swaps) which supposedly were insuring the CDOs, were at the heart of the Crash of 2007. This crash was an instance of systemic risk, in which a whole financial system collapses.

Already there are many post-mortems and best selling books and many academic studies analyzing the Crash of 2007. These autopsies are very valuable and I have read (and understood, if I might say so) most of the better ones. But I would like to advance here a few of my own perspectives.

At the Föllmer retirement celebration in Berlin, one noted speaker actually chuckled as he presented the mathematical formulas under which in principle you could gain infinite wealth. These came about by going beyond the Black–Scholes theory to the martingale (fair game betting) theory of stochastic processes. Then to price a financial instrument you must introduce utility functions which theoretically balance off how much risk you are willing to entertain in order to make a certain profit. Once this mathematical theory is in place, you attempt to optimize your gain under the risk constraints. Often this can be done by quadratic optimization of exponential expressions.

Of course, you may also lose. So you set upper bound M which is to represent the maximum amount you are willing to lose. As I learned this theory I had the uncomfortable feeling this number M was like "the tail wagging the dog." Clever stochastic mathematics to optimize your wealth ever upward would seem to imply that M was also moving upward. But that increase of M was not apparent in the clever optimization manipulations. Why not? One way this increase in potential loss could be present although not evident is hidden in the fact that you never really know what is called the Girsanov equilibrium probability measure Q. You do not know it because you do not even know the actual true probability measure P from which you must transform to the Girsanov measure Q before you bet.

My main assertion is that it was essentially a matter of human greed in allowing M to get too high in order to place a higher-yield bet so as to get a higher trade commission. Of course the investment managers were not thinking in terms of Girsanov measures! But they were indeed thinking in terms of commissions in the millions — very tempting! That my figurative M was too high was hidden in the details of the CDOs which, purported to spread risk but actually obscured it. There was no way that anyone could have monitored nor regulated the sheer complexity of such financial instruments as they were deployed in informationally inaccessible environments.

The anatomy of the Crash of 2007 which I like best is that of Gary Gorton (2008), who was the top risk mathematician for AIG. Of course there was insufficient collateral to back all those huge bets so when the collateral calls accelerated, the market collapsed, as in 1929 and 1907. Gorton emphasizes that the party could have continued longer had it not been for the introduction of the new ABX indices in 2006. These appeared in the over-the-counter market and allowed market participants to buy or sell protection against defaults of subprime bonds and mortgages. Thus, even though we could not mathematically price a CDO, nonetheless through the ABX indices humans could express their judgments about the risks. Markets are all about opinions and in 2007 the bottom fell out of the ABX market. Big alert institutions like Goldman–Sachs rushed to get their collateral calls first-in-line. Market to market accounting rules then accelerated the price declines. Large scale illiquidity by substantial financial institutions followed. Rather than letting the world financial system collapse, the important world governments stepped in with their massive financial bailouts and economic stimulus programs.

In his bestseller, Michael Lewis (2010) emphasizes that, "The beauty of the credit default swap, or CDS, was that it solved the timing problem." One did not need to guess when a crash would come. Moreover one could make the bet without laying down any cash upfront. One was leveraged to win much more than one could lose. G. Soros has called CDSs "a license to kill" and advocates banning them.

The overriding conclusion is that the human factors far outweigh the mathematical factors in explaining financial market behavior. Quants are financial engineers but traders are the salesmen and more importantly it is the CEOs and the CFOs who make the big decisions. Derivatives are here to stay but we better pay better attention to the tactics and even more importantly to the strategies of their deployments. Sometimes I use the analogy of the development of atomic weapons. Scientists were called upon to develop and understand atomic bombs and you cannot reverse the science of quantum mechanics. But it was, and is, world leaders who are responsible for the important decisions about their deployment. Warren Buffett once called derivatives "financial weapons of mass destruction." But he, and large parts of the world economy, are as locked into the use of financial derivative instruments as we all are to home and auto and health insurance. Those had to be regulated. The same attitude toward financial instruments must become a forceful part of world culture.

Such will not be easy. The analogy is the difficulty reforming the health care system in the United States. The health care industry did not want it.

That bankers will be part of the problem in reforming the financial industry is the principal message of Simon Johnson (2009). He stresses that economic "recovery will fail unless we break the financial oligarchy that is blocking essential reform."

That markets have become overly synchronized thereby increasing the likelihood and magnitude of crashes is treated in the recent book by Authers (2010). Among his recommendations for reform are the following. The megabanks must be regulated so tightly that they are not allowed to gamble, or that they must be made smaller. All the new nonbanking financial institutions must be regulated as rigorously as are banks. Limits must be imposed on the wealth that markets can create, as were enacted after the last Depression. And most importantly, somehow we must change the way we pay investment fund managers.

As to the last point, I would like to add an admittedly rhetorical but not frivolous point. Is it possible that we can educate virtually everyone in the world to understand the meaning of the mathematical concept "percentage"? Most still think that 1% is always small. . . .

Okay. Now in the next sections we go back to mathematics. If the goals in my above exposition were too large, so be it. In my mature mathematical years I have had more time to reflect on human aspects of the mathematical enterprise, e.g., see my autobiography [Gustafson (2011e)]. The recent (excellent, in my opinion) book [Patterson (2010)] also does that for the financial mathematical culture.

8.2   Quantos: Currency Options

We begin our specific operator trigonometric investigations by looking at quantos. As discussed in [Baxter and Rennie (1996)], these are financial contracts which pay off in the "wrong" currency. We consider a two-factor model with two random prices, the stock price and the exchange rate, driven respectively by two independent Brownian motions $W_1(t)$ and $W_2(t)$. Then

$$\rho W_1(t) + \sqrt{1 - \rho^2} W_2(t)$$

is also a Brownian motion with correlation ρ with the stock price Brownian motion $W_1(t)$. The stock price and the currency exchange rate follow the stochastic processes

$$S_t = S_0 \exp(\sigma_1 W_1(t) + \mu t),$$

$$C_t = C_0 \exp(\rho \sigma_2 W_1(t) + \sqrt{1 - \rho^2} \sigma_2 W_2(t) + v t),$$

where $\sigma_{1,2}$ are the respective volatilities, and μ and v the respective drifts. These μ, v drifts of course have nothing to do with our operator trigonometric entities μ_1 and v_1 but we retain the customary financial notation for stochastic drifts here. All the parameters $\sigma_{1,2}, \mu, v$, and ρ are assumed constant, volatilities $\sigma_{1,2} > 0$, $-1 \leq \rho \leq 1$, (usually $-1 < \rho < 1$), the intuition being that for large stock and currency holdings, there will be some but not perfect correlation in their movements. One also has the associated deterministic continuously compounded discount processes $B_t = \exp(rt)$ and $D_t = \exp(ut)$ where u and r are the risk-free interest rates (e.g., for treasuries). Our interest will be in the vector random variable $(\ln S_t, \ln C_t)^T$ which is assumed to be jointly normally distributed with mean $(\ln S_0 + \mu t, \ln C_0 + v t)$ and covariance matrix At, where

$$A = \begin{bmatrix} \sigma_1^2 & \rho \sigma_1 \sigma_2 \\ \rho \sigma_1 \sigma_2 & \sigma_2^2 \end{bmatrix}.$$

What is A's operator trigonometry? The quickest way to obtain A's maximum turning angle $\phi(A)$ is from A's characteristic polynomial

$$\det(A - \lambda I) = \lambda^2 - (\sigma_1^2 + \sigma_2^2)\lambda + \sigma_1^2 \sigma_2^2(1 - \rho^2)$$

from which $\lambda_1 + \lambda_2 = \sigma_1^2 + \sigma_2^2$ and $\lambda_1 \lambda_2 = \sigma_1^2 \sigma_2^2(1 - \rho^2)$, from which

$$\cos \phi(A) = \frac{2\sqrt{\lambda_1 \lambda_2}}{\lambda_1 + \lambda_2} = \frac{2\sigma_1 \sigma_2 \sqrt{1 - \rho^2}}{\sigma_1^2 + \sigma_2^2}.$$

We have immediately

Theorem 8.1 (Quantos Trigonometry). *The quantos covariance matrix operator angle $\phi(A)$ is maximal when the stock price S_t and the currency exchange rate C_t are perfectly correlated or perfectly anticorrelated. The minimum turning angle $\phi(A)$ occurs when S_t and C_t are completely uncorrelated.*

Next, to obtain the antieigenvectors $x_\pm$ of A, we need A's eigenvalues and eigenvectors. Straightforward computation gives the expressions, among which we may choose later

$$\lambda_{1,2} = \frac{1}{2}\left\{\sigma_1^2 + \sigma_2^2 \pm \sqrt{(\sigma_1^2 + \sigma_2^2)^2 - 4\sigma_1^2\sigma_2^2(1 - \rho^2)}\right\}$$

$$= \frac{1}{2}\left\{\sigma_1^2 + \sigma_2^2 \pm \sqrt{(\sigma_1^2 - \sigma_2^2)^2 + 4\sigma_1^2\sigma_2^2\rho^2)}\right\}$$

$$= \frac{1}{2}\left\{\sigma_1^2 + \sigma_2^2 \pm \sqrt{(\sigma_1^2 + \sigma_2^2)^2 - 4\sigma_1^2\sigma_2^2\sin^2\psi}\right\}$$

$$= \frac{1}{2}\left\{\sigma_1^2 + \sigma_2^2 \pm \sqrt{(\sigma_1^2 - \sigma_2^2)^2 + 4\sigma_1^2\sigma_2^2\cos^2\psi}\right\},$$

where we have taken the liberty of introducing an auxiliary angle ψ by setting $\rho = \cos\psi$ for later use. Let us call ψ the correlation angle. From $(A - \lambda I)x = 0$ we obtain the eigenvectors by the usual Gauss elimination procedure

$$x_1 = \begin{bmatrix} -2\sigma_1\sigma_2\cos\psi \\ (\sigma_1^2 - \sigma_2^2) - \sqrt{(\sigma_1^2 - \sigma_2^2)^2 + 4\sigma_1^2\sigma_2^2\cos^2\psi} \end{bmatrix},$$

$$x_2 = \begin{bmatrix} -2\sigma_1\sigma_2\cos\psi \\ (\sigma_2^2 - \sigma_1^2) + \sqrt{(\sigma_1^2 - \sigma_2^2)^2 + 4\sigma_1^2\sigma_2^2\cos^2\psi} \end{bmatrix}.$$

We may wait to normalize more advantageously, until later.

Proceeding to the weighting factors, the first weighting factor is the square root of

$$\frac{\lambda_1}{\lambda_1 + \lambda_2} = \frac{1}{2}\left\{\frac{\sigma_1^2 + \sigma_2^2 + \sqrt{(\sigma_1^2 + \sigma_2^2)^2 - 4\sigma_1^2\sigma_2^2\sin^2\psi}}{\sigma_1^2 + \sigma_2^2}\right\}$$

$$= \frac{1}{2}\left\{1 + \left[1 - \left(\frac{2\sigma_1\sigma_2}{\sigma_1^2 + \sigma_2^2}\sin\psi\right)^2\right]^{1/2}\right\}$$

$$= \frac{1}{2}\left\{1 + \left[1 - \cos^2\phi(A_0)\sin^2\psi(\rho)\right]^{1/2}\right\}.$$

Here we have inserted the operator turning angle $\phi(A_0)$ where A_0 denotes the financial situation with no correlation present, i.e., for the matrix A with $\rho = 0$. We have also reminded the reader that ψ is the ordinary trigonometric correlation angle $\psi = \psi(\rho)$ defined by $\rho = \cos \psi$ for the model we started with, with general correlation ρ.

In like manner we find the second weighting factor

$$\frac{\lambda_2}{\lambda_1 + \lambda_2} = \frac{1}{2}\{1 - [1 - \cos^2 \phi(A_0)\sin^2 \psi(\rho)]^{1/2}\}.$$

So now we have all the ingredients needed to write down the two maximally turned antieigenvectors $x_\pm$, should we need those for further analysis.

We now wish to observe an interesting geometric identity implicit within these expressions.

Lemma 8.1. *For any covariance matrix A as above,*

$$\cos \phi(A) = \cos \phi(A_0) \sin \psi(\rho).$$

Proof. See Exercise 1 and use it in the above expressions.

Lemma 8.1 trigonometrically relates the operator trigonometry of the correlated quantos model. We will use it again below, and elsewhere in this chapter.

Now that we have the operator trigonometry for this quantos model, can we see any financial interpretations that it might bring out? One may perform a change of measure to put the model into a martingale formulation. We recall that this allows you to then properly price your quanto option, assuming a no-arbitrage market. The procedure changes the drift terms to zero and results in a different stock price representation

$$S_t = S_0 \exp\left(\sigma_1 \tilde{W}_1(t) + \left(u - \rho\sigma_1\sigma_2 - \frac{1}{2}\sigma_1^2\right)t\right).$$

We ignore the discount-variance term $u - \sigma_1^2/2$ and notice the interesting drift term $-\rho\sigma_1\sigma_2$. This key term stops the dollar-discounted stock price from being a Q-martingale, with the sole exception being when the correlation value $\rho = (u - r)/\sigma_1\sigma_2$. For all other values of ρ, the dollar-valued stock is not tradable, under the usual market no-arbitrage assumption. Said another way, for all of those values, you are vulnerable to arbitrage.

Furthermore, once we have performed this (commonly called, Girsanov) change of measure to martingale representation, we may then "price up our quanto options." Letting F be the local currency forward price S_T of S_t at the maturity time T, one arrives (under some reasonable assumptions) at the quanto forward price

$$F_Q = F \exp\left(-\rho\sigma_1\sigma_2 T\right).$$

A number of interpretations then follow. For example, the quanto forward price is greater than the simple forward price iff the correlation ρ is negative, and perfect negative correlation $\rho = -1$ makes a win-win situation inevitable. In our operator trigonometric formulation of Theorem 8.1, that financial situation corresponds to maximal covariance matrix angle $\phi(A)$.

The above-found importance of the drift term $-\rho\sigma_1\sigma_2$ makes us notice that it is the first component of our eigenvectors x_1 and x_2. To sharpen the financial connection to our operator trigonometry, that suggests that we now choose to normalize x_1 and x_2 in an appropriate operator trigonometric way, notably, to further scale their expressions by $(\sigma_1^2 + \sigma_2^2)^{-1}$. Then the first eigenvector x_1 rescales to

$$x_1 = \begin{bmatrix} -\frac{2\sigma_1\sigma_2 \cos\psi}{\sigma_1^2+\sigma_2^2} \\ \frac{\sigma_1^2-\sigma_2^2}{\sigma_1^2+\sigma_2^2} - \left[\frac{(\sigma_1^2-\sigma_2^2)^2}{(\sigma_1^2+\sigma_2^2)^2} + \frac{4\sigma_1^2\sigma_2^2\cos^2\psi}{(\sigma_1^2+\sigma_2^2)^2}\right]^{1/2} \end{bmatrix}$$

$$= \begin{bmatrix} -\cos\phi(A_0)\cos\psi(\rho) \\ \sin\phi(A_0) - [1 - \cos^2\phi(A_0)\sin^2\psi(\rho)]^{1/2} \end{bmatrix}$$

$$= \begin{bmatrix} -\cos\phi(A_0)\cos\psi(\rho) \\ \sin\phi(A_0) - \sin\phi(A) \end{bmatrix}.$$

We have used Lemma 8.1. Similarly x_2 scales to

$$x_2 = \begin{bmatrix} -\cos\phi(A_0)\cos\psi(\rho) \\ \sin\phi(A_0) + [1 - \cos^2\phi(A_0)\sin^2\psi(\rho)]^{1/2} \end{bmatrix}$$

$$= \begin{bmatrix} -\cos\phi(A_0)\cos\psi(\rho) \\ \sin\phi(A_0) + \sin\phi(A) \end{bmatrix}.$$

If one then wishes to know their norms in order to divide by them to render the antieigenvectors $x_{\pm}$ both of norm one, one easily verifies, again using Lemma 8.1,

$$\|x_{1,2}\|^2 = \cos^2\phi(A_0)[\cos^2\psi(\rho) - \sin^2\psi(\rho)] + \sin^2\phi(A_0) + 1$$
$$\mp 2\sin\phi(A_0)\sin\phi(A),$$

where the $-$ goes to x_1 and the $+$ to x_2.

Let us summarize the above discussion and details in a general way as follows.

Lemma 8.2. *The key drift term $\rho\sigma_1\sigma_2$ in the quantos cross-currency pricing model, which arises in the change of measure to martingale representation, and also in the quanto forward price, has (scaled) operator trigonometric meaning*

$$\rho\sigma_1\sigma_2 \left[\frac{2}{\sigma_1^2 + \sigma_2^2} \right] = \cos\phi(A_0)\cos\psi(\rho) = \cos\phi(A)\cot\psi(\rho).$$

Here $\phi(A_0)$ is the operator angle of the uncorrelated quanto model, $\phi(A)$ is the operator angle of the full quanto model, and $\psi(\rho) = \cos^{-1}\rho$ is the scalar correlation angle. This key financial drift term also occurs in a natural way within A's eigenvectors and antieigenvectors.

Thus we see that the attainment of a no-arbitrage market condition is equivalent to exactly the "right" operator angle $\phi(A)$.

8.3   Multi-Asset Pricing: Spread Options

Spread options as general speculation devices and risk management tools are overviewed in [Carmona and Durrelman (2003)]. Generally, a spread between two asset prices $S_1(t)$ and $S_2(t)$ is the price (or value, in the case of real options such as gold mines, etc.) difference

$$S(t) = S_2(t) - S_1(t), \quad t \gtreqless 0.$$

A common example is hedging in terms of linked calls and puts.

In the following, we wish to relate spread options to our treatment of quantos in the previous section. In such a way, we will augment what we did there with a more general situation. Let two assets $S_1(t)$ and $S_2(t)$ be modeled by

$$S_i(t) = S_i(0)\exp\left[\mu_i t - \frac{1}{2}\sigma_i^2 t + \sigma_i W_i(t)\right], \quad i = 1,2.$$

Here the Brownian motions are linked by the cross-correlation expected value $E\{W_1(t)W_2(t)\} = \rho t$. The basic idea is then to discretize the mean-zero Gaussian vector $(\sigma_1 W_1(t), \sigma_2 W_2(t))^T$ with covariance matrix

$$\Sigma = \begin{bmatrix} \sigma_1^2 & \rho\sigma_1\sigma_2 \\ \rho\sigma_1\sigma_2 & \sigma_2^2 \end{bmatrix}.$$

The approach of Carmona and Durrelman (2003) is to first perform a singular value decomposition

$$\Sigma = \begin{bmatrix} \cos\theta & -\sin\theta \\ \sin\theta & \cos\theta \end{bmatrix} \begin{bmatrix} \lambda_1 & 0 \\ 0 & \lambda_2 \end{bmatrix} \begin{bmatrix} \cos\theta & -\sin\theta \\ \sin\theta & \cos\theta \end{bmatrix}^T.$$

Then one discretizes the vector $(\sigma_1 W_1(t), \sigma_2 W_2(t))^T$, using standard central differencing. The result is the interesting numerical stencil

$$\frac{1}{36} \begin{bmatrix} 1 & 4 & 1 \\ 4 & 16 & 4 \\ 1 & 4 & 1 \end{bmatrix}.$$

The precise values in this stencil come from assigning the probabilities $(1/6, 2/3, 1/6)$ to a centered difference.

What particularly interests us is the angle θ of the above singular value decomposition. The unitary matrix U in that decomposition, $\Sigma = U \begin{bmatrix} \lambda_1 & 0 \\ 0 & \lambda_2 \end{bmatrix} U^T$, strikes one as very nice indeed, since the two eigenvectors which comprise its columns are just $[\cos\theta, \sin\theta]^T$ and $[-\sin\theta, \cos\theta]^T$. So what is that angle θ, defined as

$$\theta = \arctan\left(\frac{\lambda_1 - \sigma_1^2}{\rho\sigma_1\sigma_2}\right)?$$

We note that Σ is the same as our matrix A in the quantos model. So all of our operator trigonometric analysis there applies here as well. Motivated by that previous analysis, we rescale the numerator and denominator of the θ expression to obtain the expressions

$$\tan\theta = \frac{\lambda_1 - \sigma_1^2}{\rho\sigma_1\sigma_2} = 2\left[\frac{\lambda_1 - \sigma_1^2}{\sigma_1^2 + \sigma_2^2}\right] \bigg/ \left[\frac{2\sigma_1\sigma_2\cos\psi}{\sigma_1^2 + \sigma_2^2}\right]$$

$$= \frac{\sin\phi(A) - \sin\phi(A_0)}{\cos\phi(A_0)\cos\psi(\rho)},$$

in which we have used some of our computations and results from Sec. 8.2. These new representations of $\tan\theta$ are sufficient for stating the following.

Lemma 8.3. *We may rescale our eigenvector x_1 of the quantos pricing model to be either of*

$$x_1 \sim \begin{bmatrix} 1 \\ \tan\theta \end{bmatrix} \sim \begin{bmatrix} \cos\theta \\ \sin\theta \end{bmatrix},$$

where θ is the angle of the unitary rotation matrix in the singular value decomposition from the asset spread model.

Proof. One rescales x_1 as in Sec. 8.2. The top component of x_1 there we then recognize to be the denominator in the first expression for $\tan\theta$ above. A couple of other obvious rescalings then puts $\tan\theta$ into the second component of x_1.

Of course one can always rescale a nonzero component of an eigenvector to be "anything." However, the point of Lemma 8.3 is that when we look at our expression above, we see that the (second) numerator of $\tan\theta$ is exactly the second component of our eigenvector x_1, and the (second) denominator is exactly the first component of x_1. One next asks, is eigenvector x_2 similarly related to this angle θ? But this is now clear. The matrix A satisfies $A^2 = A^T A = AA^T$ and we know the columns of U are exactly the eigenvectors of $AA^T = A^2$. Because A is symmetric, A^2 has the same eigenvectors as A. Hence the second column of U must be a rescaled x_2. In other words, up to rescalings, we must have

$$x_2 \sim \begin{bmatrix} -\sin\theta \\ \cos\theta \end{bmatrix} \sim \begin{bmatrix} 1 \\ -\cot\theta \end{bmatrix} \sim \begin{bmatrix} \tan\theta \\ -1 \end{bmatrix}.$$

Although one could verify the details of this fact by directly working from the expression for $\tan\theta$, there is a quicker way. We just compute the inner product

$$[1\ \tan\theta] \begin{bmatrix} \tan\theta \\ -1 \end{bmatrix} = 0$$

and know immediately that the new view of x_2 is true by orthogonality.

Let us summarize as follows, stating everything in terms of A, A_0, and ρ from Sec. 8.2.

Theorem 8.2 (Asset Spread). *The unitary rotation angle $\theta(U)$ of the asset spread model, the operator turning angle $\phi(A_0)$, the correlation angle $\psi(\rho)$, and the* *full operator turning angle $\phi(A)$, are in an intimate natural intertwined four-way relationship, linked by natural operator trigonometric rescalings which involve in an essential way the key change-of-measure-to-martingale drift term $\rho\sigma_1\sigma_2$.*

8.4   Portfolio Rebalancing

In [Bajeux–Besnainou and Portait (2002)] we find a study of continuous rebalancing for dynamic portfolio optimization under a mean variance criterion. The interest rate is assumed to follow a mean reverting Ornstein–Uhlenbeck process and the stock price dynamics depends on the interest rate. The setting is thus that of Markowitz portfolio risk analysis combined with Vasicek theory for risk premiums. There emerges an instantaneous variance-covariance matrix

$$\Gamma = \begin{bmatrix} \sigma_1^2 + \sigma_2^2 & \sigma_2\sigma_K \\ \sigma_2\sigma_K & \sigma_K^2 \end{bmatrix}.$$

Before we continue, let us understand financially the entries in Γ.

One has assumed a portfolio of three securities: stocks, bonds, and cash. The stock fund is assumed to follow a stochastic process

$$\frac{dS_t}{S_t} = (r(t) + \theta_s)dt + \sigma_1 dz + \sigma_2 dz_r$$

where $r(t)$ is an instantaneous stochastic riskless interest rate process

$$dr(t) = a_r(b_r - r(t))dt - \sigma_r dz_r.$$

There dz and dz_r are orthogonal standard Brownian motions. The bond fund follows a stochastic process

$$\frac{dB_t}{B_t} = (r(t) + \theta_K)dt + \sigma_K dz_r$$

which is assumed to be continuously rebalanced to maintain a constant maturity date K over the period $0 \stackrel{\leq}{=} t \stackrel{\leq}{=} T$. The cash fund is essentially just a time value of money process, e.g., a money market fund

$$\frac{dM_t}{M_t} = r(t)dt.$$

The θ_S and θ_K are constant risk premiums of the stock and bond funds, respectively.

We are going to focus primarily on the matrix Γ. The σ_1 and σ_2 volatilities come from the stock fund and the σ_K volatility comes from the bond fund. The σ_r volatility of the interest rate process does not appear directly in Γ because it is related to σ_K and K according to Vasicek theory and by an assumption of constant market price of stock market risk.

Γ is a symmetric positive definite matrix. Let us inquire into its operator trigonometry. From its characteristic polynomial

$$\det(\Gamma - \lambda I) = \lambda^2 - (\sigma_1^2 + \sigma_2^2 + \sigma_K^2)\lambda + \sigma_1^2\sigma_K^2$$

we have immediately

$$\cos\phi(\Gamma) = \frac{2\sigma_1\sigma_K}{\sigma_1^2 + \sigma_K^2 + \sigma_2^2}$$

which gives us Γ's operator trigonometric maximal turning angle $\phi(\Gamma)$. Straightforward computation gives Γ's eigenvalues

$$
\begin{aligned}
\lambda_{1,2} &= \frac{(\sigma_1^2 + \sigma_K^2 + \sigma_2^2)}{2}\left\{1 \pm \left[1 - \frac{4\sigma_1^2\sigma_K^4}{(\sigma_1^2 + \sigma_K^2 + \sigma_2^2)^2}\right]^{1/2}\right\} \\
&= \frac{1}{2}\mathrm{tr}(\Gamma)\left\{1 \pm \left[1 - \cos^2\phi(\Gamma)\right]^{1/2}\right\} \\
&= \frac{1}{2}\mathrm{tr}(\Gamma)\left\{1 \pm \sin\phi(\Gamma)\right\}.
\end{aligned}
$$

Notice that the last expression also corresponds to obtaining $\sin\phi(\Gamma)$ from its characterization as eigenvalue spread divided by eigenvalue sum. We may also write $\sin\phi(\Gamma)$ in a complete "trace-like" form

$$\sin^2\phi(\Gamma) = \frac{((\sigma_1 - \sigma_K)^2 + \sigma_2^2)((\sigma_1 + \sigma_K)^2 + \sigma_2^2)}{(\sigma_1^2 + \sigma_K^2 + \sigma_2^2)^2}.$$

Continuing, for the eigenvectors, direct computation as in Sec. 8.2 gives

$$x_{1,2} = \begin{bmatrix} -2\sigma_2\sigma_K \\ (\sigma_K^2 - \sigma_1^2 - \sigma_2^2) \mp [(\sigma_1^2 + \sigma_2^2 + \sigma_K^2)^2 - 4\sigma_1^2\sigma_K^2]^{1/2} \end{bmatrix}.$$

We have used the same scaling trick after the Gauss procedure that we used in Sec. 8.2, i.e., one lets the first component be $-2\sigma_2\sigma_K$. From the above ingredients we also have in principle the two antieigenvectors $x_\pm$.

We observe that the operator trigonometry of Γ is a bit more complicated than that of A and Σ. The $\sigma_2\sigma_K$ first component of the eigenvectors is not immediately compatible with the operator cosine. Generally, we can say that the stochastic couplings of the stock and bond portfolio processes to the interest rate process via the volatilities σ_1 and σ_K appear to create this more complicated operator trigonometric turning geometry.

Because the volatility σ_K importantly couples bond yields to the interest rates, as it should, it would appear to be the more essential ingredient of the model. If we would like to simplify the model, then one might think of dropping the σ_2 coupling to interest rate stochasticity in the stock price process. This makes the operator cosine simpler, but then Γ is diagonal with obvious operator trigonometry.

Let us delve a little deeper into this portfolio rebalancing under stochastic interest rates. A main financial point is to optimize the portfolio according to the criteria of maximizing logarithmic utility. This leads to the need to solve the linear system $h = \Gamma^{-1}\theta$ for optimal weights h. This system and its solution are (we have consolidated some details here)

$$\begin{bmatrix} h_S \\ h_K \end{bmatrix} = \begin{bmatrix} \frac{1}{\sigma_1^2} & \frac{-\sigma_2}{\sigma_K\sigma_1^2} \\ \frac{-\sigma_2}{\sigma_K\sigma_1^2} & \frac{\sigma_1^2+\sigma_2^2}{\sigma_K^2\sigma_1^2} \end{bmatrix} \begin{bmatrix} \theta_S = \sigma_1\lambda + \sigma_2\lambda_r \\ \theta_K = \sigma_K\lambda_r \end{bmatrix} = \begin{bmatrix} \frac{\lambda}{\sigma_1} \\ \frac{-\lambda\sigma_2+\lambda_r\sigma_1}{\sigma_1\sigma_K} \end{bmatrix}.$$

The drift parameter θ_K is the risk premium of the bond fund. The drift parameter θ_S is the risk premium of the stock fund. Both depend fundamentally on the market price of interest rate risk λ_r, and θ_S depends also on the market price of stock market risk.

There does not seem much more for us to do here, but we do feel it is worthwhile to make the following point. Should one wish to further analyze the system operator-trigonometrically to determine the geometric behavior of solutions, e.g., as related to the matrix Γ^{-1}, the operator trigonometry is already available from our considerations above.

8.5   American Options with Random Volatility

The American put option exercise boundary is treated in [Touzi (1999)] under the more realistic assumption of nonconstant volatility parameter. For American options, in which one can choose to exercise the option at any time up to the expiry date T, a free boundary problem appears. This boundary is called the exercise boundary. Generally, it is not known in explicit analytic form, but a key notion is that of the high-contact (also called: smooth-fit) condition: the delta $\delta V/\delta S$ must be -1 at this free boundary. Here S is the asset price and V is the American put option value. Such regularity issues of the exercise boundary when the volatility parameter is itself a stochastic process are studied in this paper.

The stochastic price process S_t for the asset and the random volatility process σ_t are coupled via the stochastic differential system

$$\frac{dS_t}{S_t} = \mu(t, S_t, Y_t)dt + \sigma(Y_t)[\sqrt{1 - \rho^2(t, Y_t)}dW_t^S + \rho(t, Y_t)dW_t^\sigma,$$

$$dY_t = \eta(t, Y_t)dt + \gamma(t, Y_t)dW_t^\sigma,$$

where the W_t are Brownian motions, $\sigma_t = \sigma(Y_t)$ is the Markov volatility process of the asset S_t, $\rho_t = \rho(t, Y_t)$ is the instantaneous correlation process between S_t and σ_t, $-1 \leq \rho_t \leq 1$, and ρ_t is assumed to be independent of the asset price S_t. From these and some other technical assumptions, one writes down the dispersion matrix

$$\Sigma_t = \begin{bmatrix} \sigma(Y_t)\sqrt{1 - \rho^2(t, Y_t)} & \sigma(Y_t)\rho(t, Y_t) \\ 0 & \gamma(t, Y_t) \end{bmatrix},$$

from which one is led to the symmetric positive definite matrix

$$A_t = \Sigma_t \Sigma_t^* = \begin{bmatrix} \sigma^2 & \gamma\sigma\rho \\ \gamma\sigma\rho & \gamma^2 \end{bmatrix},$$

in which we have suppressed the dependencies on $Y(t)$ and t. The μ and η drift terms disappear under a Girsanov change to equivalent martingale measure, and by a change of variable $x = \ln S$, one arrives at a uniformly elliptic partial differential operator with the A_t matrix as its coefficient matrix. This is the key step in proving the high-contact condition.

Here, we may turn our interest to the uniformly elliptic partial differential equation coefficient matrix. Immediately we have its operator trigonometry, by comparison to the matrix A of Sec. 8.2. By that comparison, namely,

$$\begin{bmatrix} \sigma^2 & \rho\sigma\gamma \\ \rho\sigma\gamma & \gamma^2 \end{bmatrix} \sim \begin{bmatrix} \sigma_1^2 & \rho\sigma_1\sigma_2 \\ \rho\sigma_1\sigma_2 & \sigma_2^2 \end{bmatrix},$$

one may write down the full operator trigonometry of this section from that
of Sec. 8.2, via the symbol replacement $\gamma \sim \sigma_2$. Of course the parameter,
stochastic, and financial model meanings are quite different.

8.6   Risk Measures for Incomplete Markets

In this section we consider the important recent paper [Carr, Geman,
Madan (2001)] in which a new approach for pricing and hedging in
incomplete markets is proposed and compared to other recent attempts
to get good models for the incomplete market situation. Here, we are ven-
turing into somewhat uncharted territory, both with respect to financial
instrument theory and the operator trigonometry. As to the former, it is
now recognized that the usual assumptions of complete markets and the
no-arbitrage assumption are suspect and do not really hold. Thus the study
of risk measures for incomplete markets is highly topical in the financial
engineering community. As to the operator trigonometry, it is fully developed for symmetric positive definite matrices, but less so for general matrices. Thus, for both reasons, the reader may regard this section of this chapter
on the operator trigonometry of multi-asset financial models as incomplete
and preliminary. It may be viewed as an attempt to encounter and see what
might be needed in the future. Because the paper [Carr *et al.* (2001)] touches
on many current financial modeling issues, we will here just extract one
incomplete market model from it.

Without further ado, let us then go to the example given there on
p. 157. The setting is a single-period continuous-state economy with log-
normal valuation test measures. One has a stock and a bond and a maturity date T. The stock currently has value S_0 and the unit bond has current
value (i.e., already continuously discounted) e^{-rT}. One tries two valuation
test measures. The first has effective rate of return $\mu_d < r$ (below market)
and volatility $\sigma_d > 0$. The second has rate of return $\mu_u > r$ (above market)
and volatility $\sigma_u > \sigma_d$. The first test measure represents an investor who
evaluates wealth in dollars, the second an investor who measures wealth
in stock shares. After a few more risk modeling details, one arrives at a 2×2
matrix of asset valuation test measure expectation-based outcomes

$$C = \begin{bmatrix} S_0 e^{(\mu_u - r)T} & S_0 e^{(\mu_d - r)T} \\ e^{-rT} & e^{-rT} \end{bmatrix}.$$

Consider a zero cost trading strategy $\alpha = (\alpha_0, \alpha_1)$ with an initial portfolio balance

$$\alpha_0 S_0 + \alpha_1 e^{-rT} = 0.$$

Then the product $\alpha^T C$ is

$$\alpha^T C = \alpha_0 S_0 (e^{(\mu_u - r)T} - 1, \ e^{(\mu_d - r)T} - 1).$$

Because $\mu_u > r > \mu_d$, this payoff cannot be positive, no matter how you choose your zero cost strategy. So the desired NSAO (no strictly acceptable opportunities) condition holds. This and the fact the C is invertible guarantees the existence of a unique representative pricing function (RSPF), which is then calculated. The result is a strategy $\alpha^* = (\alpha_0^*, \alpha_1^*)$, the calculation of which requires two Black–Scholes subcalculations. The resulting strategy is called "just acceptable."

The first point we wish to make here is that our operator trigonometry, as presently constituted, does not yet directly apply to the outcome matrix C. Our theory applies to positive definite matrices, be they symmetric, or more generally, with positive definite real part. The matrix C exhibits the other kind of matrix positivity, what one calls a positive cone map, or you may view it more specifically as a matrix positive in the Perron sense: all the entries $c_{ij} > 0$.

I do have an operator trigonometry [Gustafson (2000d)] for arbitrary invertible matrices, based upon the polar form representation such matrices have: $A = U|A|$, where U is a unitary map and $|A|$ is the absolute value operator for A, i.e., $|A| = (A^*A)^{1/2}$. The resulting operator trigonometry for A is then essentially that of the symmetric positive definite operator $|A|$. Let us compute C^*C for the outcome matrix C. We find

$$C^*C = e^{-2rT} \begin{bmatrix} S_0^2 e^{2\mu_u T} + 1 & S_0^2 e^{(\mu_u + \mu_d)T} + 1 \\ S_0^2 e^{(\mu_u + \mu_d)T} + 1 & S_0^2 e^{2\mu_d T} + 1 \end{bmatrix},$$

or

$$e^{2rT} C^*C = 2 \begin{bmatrix} 1/2 & 1/2 \\ 1/2 & 1/2 \end{bmatrix} + S_0^2 \begin{bmatrix} (e^{\mu_u T})^2 & e^{\mu_u t} e^{\mu_d T} \\ e^{\mu_u T} e^{\mu_d T} & (e^{\mu_d T})^2 \end{bmatrix}.$$

We wrote the latter to bring out the first term, which singles out the Perron Projector onto the span of the positive Perron equiprobability vector, namely,

$$P = \frac{pp^T}{p^Tp} = \begin{bmatrix} 1/2 & 1/2 \\ 1/2 & 1/2 \end{bmatrix}, \quad p = [1/2 \quad 1/2]^T.$$

P is the rank-one orthogonal projector onto the span of p. Although not an operator trigonometric result, it may be of interest for the treatment of similar incomplete-market expectation-based outcome models. It represents a decomposition of the (highly incomplete) market model into stochastic and drift components.

One could now analyze this model operator trigonometrically. For example, one could calculate $|C|$ and thereby get the polar decomposition $C = U|C|$ and then impose the extended operator trigonometry.

Commentary

This is a first inquiry into operator trigonometries of financial instruments. It has revealed some interesting financial operator trigonometries when the model matrices are symmetric positive definite, or if one is willing to go to symmetrized versions such as singular value decompositions. The operator trigonometry as presently constituted is not yet ready for application to general linear models.

In this chapter we focused on two-factor models, which bring out the main issues, results, and interpretations. However, one may in the same way treat n-dimensional multi-factor financial instruments and models, provided they be formulated in terms of covariance-like matrices, or more generally, symmetric positive definite matrices. Then one will also have the higher antieigenvalues, and resulting critical (decreasing in size) turning angles $\phi_k(A)$. As seen in Chapter 6, these entities play important roles in the applications of the operator trigonometry to statistics, and we would also expect them to do so in multivariate finance.

Most financial instrument pricing theory is based upon either utility theory, Black–Scholes theory, the more recent no-arbitrage theory, or combinations thereof. The currency pricing quantos problem we treated in Sec. 8.2 was based on no-arbitrage theory. The spread option pricing problem we treated in Sec. 8.3 and the random volatility problem we treated in Sec. 8.5 were based on Black–Scholes theory. The optimal growth portfolio problem we treated in Sec. 8.4 was based on maximizing logarithmic utility. The risk measures analysis we treated in Sec. 8.6 concerned pricing-hedging models intermediate between utility theory and no-arbitrage pricing theory.

Recently I have looked at financial portfolio discussions suggesting that, although riskier, one should use the geometric mean rather than the arithmetic mean in the important Sharpe ratio. Immediately I may interpret [Gustafson (2012a)] the ratio (with the geometric mean on top) of those ratios within my antieigenvalue theory. Increasing risk corresponds in my new ratio of ratios to decreasing antieigenvalue and thus to widening angle (greater growth potential) between the portfolios.

8.7   Exercises

1. It always surprises me how some new operator trigonometric fact can jump out when one looks at a new application. The following were useful in treating financial instruments. Verify them.

   Lemma 8.4. (a) *The weighting factors in the antieigenvectors $x_\pm$ may be expressed in terms of the operator turning angle as follows:*

   $$\frac{\lambda_1}{\lambda_1 + \lambda_n} = \frac{1 + \sin\phi(A)}{2}, \quad \frac{\lambda_n}{\lambda_1 + \lambda_n} = \frac{1 - \sin\phi(A)}{2}.$$

   (b) *The antieigenvectors of A^{-1} are*

   $$x_\pm(A^{-1}) = \pm\left(\frac{\lambda_1}{\lambda_1 + \lambda_n}\right)^{1/2} x_1 + \left(\frac{\lambda_n}{\lambda_1 + \lambda_n}\right)^{1/2} x_n.$$

2. Verify the proof of Lemma 8.1 in Sec. 8.2.
3. Verify Lemma 8.2, that the key financial drift term $\rho\sigma_1\sigma_2$ has (scaled) operator trigonometric meaning

   $$\rho\sigma_1\sigma_2\left[\frac{2}{\sigma_1^2 + \sigma_2^2}\right] = \cos\phi(A_0)\cos\psi(\rho) = \cos\phi(A)\cot\psi(\rho),$$

   where A_0 is the matrix with correlation ρ taken to zero.
4. Verify the $h = \Gamma^{-1}\theta$ calculation finding the optimal portfolio rebalancing weights in Sec. 8.4.
5. Learn about, and derive, the Black–Scholes PDE.
6. E. Thorp is credited with being the "godfather of the quants" in the book by Patterson (2010). Find an early paper of mine where Thorp and I were competing in the same pure mathematics.

Chapter 9

Other Directions

Perspective

Of course I expect further development of the antieigenvalue theory and operator trigonometry within the contexts of the previous Chapters 2 through 8. However, the purpose of this monograph was to assemble the theory and applications as I know them, and to now pass the tasks and opportunities for further development to others.

How, then, to conclude the book? I decided to be brief. Any attempt at a systematic treatment of envisioned new results would be self-defeating because such would take years. Therefore to enable brevity within some sense of comprehensiveness, I organized the content of this last chapter to be (in sections): 9.1 Operators; 9.2 Angles; 9.3 Optimization; 9.4 Equalities; 9.5 Geometry; 9.6 Applications. What you will find therein will be generalizations, speculations, loose ends, and above all . . . incompleteness.

9.1 Operators

My own preference in creating this theory of antieigenvalues and its associated operator trigonometry has clearly been to do so for positive definite symmetric, Hermitian, thus generally selfadjoint operators. In retrospect, there are two reasons for this. First, that is where the theory is most complete and the turning angles most visible. Second, that is where the applications took me.

One naturally asks, however, how about more general operators, and specifically, how about more general matrices? In [Gustafson (2000d)] I extended the theory to all invertible matrices A via its positive definite absolute value $|A|$. One could go to noninvertible A and utilize all the generalized inverses A^- so useful in statistics and group theory, but I have not done that.

One guideline as to what realistically should be attempted has been the Rayleigh–Ritz variational theory for eigenvalues. One likes real eigenvalues for physical reasons, and diagonalizable matrices for mathematical reasons. Notice that the Rayleigh–Ritz theory has been essentially limited to selfadjoint operators. Some of it goes through for normal operators. Beyond those, there is the famous attempt by Dunford and Schwartz (1971) in their third volume to go beyond normal operators to what they called spectral operators. Not much came out.

C. Davis (1980) and my former Ph.D. student M. Seddighin took the lead in taking the antieigenvalue theory to normal matrices and compact normal operators. Apart from what is found in the books [Gustafson (1997a, d)] and our joint papers, see also Morteza's papers [Seddighin (2002; 2005; 2009; 2011)] and his citations therein. I have decided not to go into that somewhat more technical theory here. In a sense, I am waiting for more applications.

However, I would like to expose here the not widely appreciated fact, that all diagonalizable matrices A can be made normal provided that one is allowed to re-inner-product the space. Such tactic is quite useful in treating recurrence lengths for minimum residual iterative schemes currently employed for very large linear systems $Ax = b$. See for example [Axelsson (1994); Gustafson (2004b)]. Here is how it goes.

Let us consider any diagonalizable matrix A. It has n linearly independent eigenvectors and is diagonalized according to

$$S^{-1}AS = D = \begin{bmatrix} \lambda_1 & & \\ & \ddots & \\ & & \lambda_n \end{bmatrix}$$

better seen via the intertwining

$$AS = SD,$$

from which we know the columns of S are exactly A's eigenvectors. A may then be made a normal operator in the inner product

$$\langle x, y \rangle_B = \langle Bx, y \rangle,$$

where $B = (SS^*)^{-1}$ is SPD in the original inner product. This follows from the fact (see Exercise 1) that in the new inner product space any matrix A has dual

$$A' = B^{-1}A^*B.$$

To see that A is then normal in the new inner product, we first compute

$$A' = SS^*A^*(S^*)^{-1}S^{-1}$$

$$= SS^*[(S^{-1})^*D^*S^*](S^*)^{-1}S^{-1}$$

$$= SD^*S^{-1}.$$

Then we have

$$A'A = (SD^*S^{-1})(SDS^{-1})$$

$$= SD^*DS^{-1}$$

and

$$AA' = (SDS^{-1})(SD^*S^{-1})$$

$$= SDD^*S^{-1}.$$

But $D^*D = DD^*$ so $A'A = AA'$.

Thus diagonalizable is equivalent to normalizable in this context. Also in the new inner product one finds the convenient relation

$$\langle Ax, x \rangle_B = \langle BAx, x \rangle = \langle (S^*)^{-1}S^{-1}Ax, x \rangle$$

$$= \langle S^{-1}Ax, S^{-1}Ax \rangle = \langle DS^{-1}x, S^{-1}x \rangle$$

$$= \langle Dy, y \rangle$$

where $y = S^{-1}x$.

Thus one could develop the full antieigenvalue theory and operator trigonometry for A in this new inner product. See Exercise 1 (answers) for an example.

There are many, many ways one may extend the operator trigonometry to other operators in other settings, for example to semi-inner product Banach spaces. Here there are many loose ends. I would suppose a good criterion for going in those directions would be to require new applications to guide such investigations. Equally valid would be some nice theoretical insight. But I will say no more here.

9.2 Angles

I have preferred staying with operator turning angles $\phi(A)$ that I can actually see. Since they depend only on antieigenvector x being turned $\phi(A)$ degrees when A maps x to Ax, one only needs to see the trigonometry within that particular two-space sp$\{x, Ax\}$. As we have seen in this book, $\phi(A)$ is often better thought of intuitively as a maximal twisting, rather than turning, action by A. We have also seen that for general A written in polar form $A = U|A|$, one should not confuse my operator trigonometry with uniform rotation angles that might be resident within the unitary factor U, nor with phase angles in a nonreal eigenvalue $\lambda = re^{i\psi}$. Generally, and that is the bottom line of this section, I have not had much success in developing an operator trigonometry involving complex-valued angles. To do so would open new vistas.

Of course one can do it formally. That is why I originally in [Gustafson (1972a)] defined an imaginary operator trigonometry and the total operator trigonometry. The thinking then was just to do it variationally, in a way analogous to the variational Rayleigh–Ritz theory. As the examples in [Gustafson (2000d)] show, one does not get much out of such theory, at least not yet.

Probably one will eventually have to deal better with phase, and I would prefer that such be investigated from some real physical applications where it is really needed.

As a generalization of my [Gustafson (1972)] imaginary and total antieigenvalue concepts, recently Hossein, Paul, Debnath and Das (2008) introduced a notion of symmetric antieigenvalue. See also recent papers by these authors for their versions of my theory. Being perhaps a bit proprietary, I did not see any new breakthroughs in most of those papers, although on the other side, as referee I accepted some of them to guarantee publication of their sometimes interesting variations on the antieigenvalue theory.

However, we did find the just-mentioned symmetric antieigenvalue paper susceptible to a much wider generalization and at the same time reducible to my original antieigenvalue concepts, see [Gustafson (2010b)]. In that paper Seddighin and I defined what we called slant antieigenvalues and corresponding slant antieigenvectors. These include all the total, imaginary, and symmetric antieigenvalue concepts. Then, more importantly, we showed how they all can be reduced conceptually to my original (real) antieigenvalue concept.

Consider, for example, the symmetric antieigenvalue angle defined variationally in terms of

$$\phi(f) = \frac{\text{Re}\,\langle Tf,f\rangle + \text{Im}\,\langle Tf,f\rangle}{\sqrt{2}\|Tf\|\|f\|}.$$

Recalling the decomposition $T = \text{Re}\,T + i\,\text{Im}\,T$ of a bounded operator T on a Hilbert space, we see that for $\|f\| = 1$,

$$\begin{aligned}
\phi(f) &= \frac{\langle(\text{Re}\,T)f,f\rangle + \langle(\text{Im}\,T)f,f\rangle}{\sqrt{2}\langle Tf,Tf\rangle^{1/2}} \\
&= \frac{1}{2i}\frac{\langle(iT + iT^* + T - T^*)f,f\rangle}{[2\langle T^*tf,f\rangle]^{1/2}} \\
&= \frac{1}{2\sqrt{2}}\frac{\langle[(1-i)T + (1+i)T^*]f,f\rangle}{\|Tf\|}.
\end{aligned}$$

Notice now that $[(1-i)T]^* = (1+i)T^*$, and accordingly, if I let

$$A = (1-i)T,$$

then the above symmetric antieigenvalue expression becomes

$$\begin{aligned}
\phi(f) &= \frac{1}{2}\frac{\langle(A + A^*)f,f\rangle}{\langle A^*Af,f\rangle^{1/2}} \\
&= \frac{\langle(\text{Re}\,A)f,f\rangle}{\|Af\|\{f\}} \\
&= \text{Re}\,\frac{\langle Af,f\rangle}{\|Af\|\|f\|}.
\end{aligned}$$

Thus, by a 45° rotation, we are back to my original 1972 theory.

Our more general slant antieigenvalue theory allows arbitrary angle rather than the 45° angle implicit in the symmetric antieigenvalue theory. We show in [Gustafson (2010b)] how to also bring our generalization back to my original real eigenvalue concept. Generally I must criticize our paper and also the symmetric antieigenvalue theory as not driven at all by any compelling physical application. But perhaps such will show up in the future.

Another mathematical generalization has been to what I call interaction-antieigenvalues. I mentioned this in my 1972 paper and finally got back to it in [Gustafson (2004c)]. I found interesting new connections of this concept to key inequalities developed independently in other contexts by [Strang (1962); Stampfli (1970); Greub and Rheinboldt (1959)]. See also [Seddighin (2005)] for how interaction antieigenvalues play out in normal subalgebras.

My early result for the first antieigenvalue μ of an SPD matrix A, that

$$\mu = \cos \phi(A) = \frac{2\sqrt{\lambda_1 \lambda_n}}{\lambda_1 + \lambda_n},$$

can be seen as a ratio of means:

$$\mu = \frac{\text{geometric mean}}{\text{arithmetic mean}}.$$

Clearly one could envision a whole antieigenvalue theory in terms of ratios of means of various kinds. But would such really be geometrical? That needs to be further explored.

Kaporin (1994) independently valuably used such ratios as condition numbers for iterative schemes in computational linear algebra. His condition number was defined to be

$$\beta(A) = \left(\sum_{i=1}^{n} \frac{\lambda_i}{n} \right)^n \bigg/ \prod_{i=1}^{n} \lambda_i.$$

The connection to my anti-eigenvalue is immediate if you take $n = 2$: then $\beta^{-1/2} = \mu$. Their general relationship and to the usual condition number κ is

$$1 \leqq \beta(A)^{1/n} \leqq \kappa(A) \leqq [\kappa(A)^{1/2} + \kappa(A)^{-1/2}]^2 = 4\mu(A)^{-2} \leqq 4\beta(A).$$

For more on such connection, see [Gustafson (1999d)].

Recall that in Sec. 4.6 I emphasized the merit of thinking of my antieigenvalue μ as a condition number. The usual condition number $\kappa(A) = \sigma_1(A)/\sigma_n(A)$ in terms of the eigenvalues of $|A|$ measures relative stretching actions by A along principal axes. My $\mu = \cos \phi(A)$ and the higher antieigenvalues μ_{ij} measure relative twisting actions by A between principal axes. Creating an antieigenvalue theory according to ratios of means would resemble what we have already encountered in Sec. 6.4 where I could treat operator-trigonometrically inequalities from matrix statistics.

When you statistically average, you lose individual precision. On the other side, your various statistical means at times carry more validity than your individual observations. So I am neutral about such a future antieigenvalue theory based upon every statistical mean you could put your hand on. But it could be mathematically interesting.

Somewhat as an afterthought here, perhaps I should mention that to my knowledge no one has looked further into a theory of the 90° turning angle. Recall how I mentioned in Sec. 2.4 that very early on I was unhappy to find that the cosine of unbounded accretive operators in a Hilbert space is zero. Later, in Lemma 2.1 I showed how you could make those unbounded operators into bounded ones by using the graph norm. But I have not gone further to look into any possibly useful features of an operator trigonometry within such graph norms. For example, A will remain accretive in the new inner product, and m-accretive also if m-accretive originally. But now A is bounded. So my original theory for turning angles less than 90° comes back into relevance.

In that connection I should mention that our paper [Gustafson (1969b)] was extended by P. Hess (1971) as follows. Let T be an unbounded linear map from normed linear space X into normed linear space Y. Let $Q(x, y)$ be a sesquilinear form satisfying $|Q(x,y)| \overset{\leq}{=} \|x\|\|y\|$ on $X \times Y$. Define $\cos_Q(T) = \inf |Q(x, Tx)|/\|Tx\|\|x\|$ for all x in the domain of T, $x \neq 0$, $Tx \neq 0$. Then $\cos_Q(T)$ is zero. I had briefly known Hess as a promising young Swiss mathematician. Unfortunately he died at a relatively young age while hiking in Utah.

9.3 Optimization

We have seen in Theorem 3.1 in Sec. 3.2 that the norm convex minimum that yields $\sin \phi(A)$ constitutes, in my view, a rather general optimization criteria. That is, it encompasses all vectors x in the joint unit sphere of the span of the smallest and largest eigenspaces. Moreover $\sin \phi_{ij}(A)$ performs as an optimizor in the same way for all the higher and skew turning angles of A, as we saw, for example, in Sec. 6.4. Therefore I would like to advertise these $\sin \phi_{ij}(A)$ convex minima as worthy of being embedded into the general culture of optimization. Much of optimization theory deals with convex minima, and many cost functions are quadratic. Those are also the context attributes of my $\sin \phi(A)$.

If I have time, I would not mind advancing this fancy of mine myself in the future. But I have not had time yet to look into details. Therefore in this section I hope to just provide a glimpse of what might be out there.

We have already seen precursors in Sec. 6.2. See Theorem 6.2. The inefficiency vectors

$$x^{j,k} = \frac{1}{\sqrt{2}}x_j \pm \frac{1}{\sqrt{2}}x_k$$

have equal weights $2^{-1/2}$, whereas the antieigenvectors have their special counterweights. I say special because it is my expectation that in the future it will turn out that the odds within my envisioned optimization venue will favor equal weights in the majority of applications.

Here is an example. In Sec. 4.1 I mentioned the Kantorovich–Wielandt inequality

$$\frac{|\langle Ax, Ay \rangle|}{\|Ax\| \|Ay\|} \stackrel{\leq}{=} \cos \theta,$$

where the angle θ was defined in terms of the matrix condition number κ according to $\cot(\theta/2) = \kappa$. The constraint on x and y was that they must be an orthonormal pair of vectors. It is easy to show that the optimizing K–W inequality vectors are

$$x = x_1 + x_n, \quad y = x_1 - x_n,$$

i.e., they are equally weighted.

Even though optimization with equal weights may prove to be more prevalent, optimization with unequal weights will be more interesting.

Here is an example with unequal weights. The Shisha–Mond inequality (see, e.g., [Drury *et al.* (2002)]) says that

$$\max_{\|x\|=1} \left[\langle Ax, x \rangle - \frac{1}{\langle A^{-1}x, x \rangle} \right] \stackrel{\leq}{=} (\sqrt{\lambda_1} - \sqrt{\lambda_n})^2.$$

This is optimized at

$$x = \left(\frac{\lambda_1^{1/2}}{\lambda_1^{1/2} + \lambda_n^{1/2}} \right)^{1/2} x_1 \pm \left(\frac{\lambda_n^{1/2}}{\lambda_1^{1/2} + \lambda_n^{1/2}} \right)^{1/2} x_n.$$

I leave it to the reader in Exercise 4 to verify that such x do indeed optimize the inequality.

A pause for reflection reveals an interesting wrinkle. The weights are unequal, but they resemble my antieigenvector weights. They do not, however, counterweight A's eigenvectors. But through use of my Lemma 8.4 from Exercise 1 of Chapter 8, they are exactly the weights of the antieigenvectors of $A^{-1/2}$. In other words, somehow the Shisha–Mond inequality is within the confines of my antieigenvector theory. This observation is new here. In Exercise 4 I suggest the reader inquire further.

But I must move on.

9.4   Equalities

In Sec. 7.2 I went beyond the Bell inequalities by creating, from them, what I called Bell equalities. This fills in probability and physical gaps in our understanding of such quantum measurement theory and all the attendant issues. Then in Exercise 2 of Chapter 7 I proposed the very general program in mathematics of attempting, where possible, to render all inequalities into equalities. In the answers to that exercise, I trotted out the trivial example of the Cauchy–Schwarz inequality made equality. Here I would like to elaborate with some more observations on this theme.

My Min-Max Theorem (Theorem 1.1) is actually itself an example. The left-hand side is always less than or equal to the right-hand side. Only the antieigenvectors can bring it up to equality.

But such a situation could just be viewed as the issue of when an inequality can be made sharp.

What I propose is more: to be able to create analogs of the "slack variables" that one finds, for example, in linear programming theory, e.g., see [Luenberger (1973)]. Let me recall that situation there. You need to minimize some cost function, often just a linear combination of the independent variables, subject to a system of linear inequality constraints. So you fill in those linear inequality constraints with new linear variables "to take up the slack," so to speak.

Going back to my Min-Max Theorem, in a weak sense one may argue that my constructive proof did create approximating conic sections which had to sweep upward through a "slack feasibility region" below the sought convex curve $\|\epsilon A - I\|^2$. In fact it could be interesting to look more closely at a quadratic theory of slack variables. For example, if you look at my proof of Theorem 3.1 in Sec. 3.2, there I expand the left-hand side of the Min-Max Theorem into the expression

$$\max_{\|x\|=1} \min_{\epsilon>0} \|(\epsilon A - I)x\|^2 = \max_{\|x\|=1} \left[1 - \left(\frac{\operatorname{Re} \langle Ax, x \rangle}{\|Ax\|} \right)^2 \right].$$

The $(\mathrm{Re}\,\langle Ax, x\rangle / \|Ax\|)^2$ could be interpreted as a slack variable which must be minimized by bringing x to an antieigenvector.

However, and I wish to belabor this point, what I proposed in Exercise 2 of Chapter 7 is a stronger mandate: to bring my approach to all of mathematical inequality theory, exemplified by the rendering of the important Bell inequalities into new quantum spin identities. In [Gustafson (2003c)] I coined the term *inequality–equalities*. To emphasize what I intend, let us recall one of the simplest Bell inequalities from Sec. 7.2, the Wigner inequality

$$\frac{1}{2}\sin^2\left(\frac{\theta_{23}}{2}\right) + \frac{1}{2}\sin^2\left(\frac{\theta_{12}}{2}\right) \geq \frac{1}{2}\sin^2\left(\frac{\theta_{31}}{2}\right).$$

This became the inequality–equality

$$\sin^2\left(\frac{\theta_{23}}{2}\right) + \sin^2\left(\frac{\theta_{12}}{2}\right) - \sin^2\left(\frac{\theta_{13}}{2}\right)$$

$$= 2\cos\left(\frac{\theta_{13}}{2}\right)\left[\cos\left(\frac{\theta_{13}}{2}\right) - \cos\left(\frac{\theta_{12}}{2}\right)\cos\left(\frac{\theta_{23}}{2}\right)\right].$$

The right-hand side took up the slack, and sharply. As it turned out, the slack variables happened to be all cosines. That is interesting quantum mechanically.

Let us therefore go back to kindergarten and that the mundane Cauchy–Schwarz inequality be made equality

$$|\langle u, v\rangle| = \|u\|\,\|v\|\cos\theta_{u,v}.$$

Although I would be the first to assert this as the most important inequality–equality in mathematics, can one do something more impressive? It was such thinking that led me in [Gustafson (2001a)] to a new formulation and meaning of $\sin\phi(A)$.

The key to this new formulation of $\sin\phi(A)$ is to start with the Lagrange identity

$$\left(\sum_{i=1}^{n} a_i^2\right)\left(\sum_{i=1}^{n} b_i^2\right) - \left(\sum_{i=1}^{n} a_i b_i\right)^2 = \sum_{1\leq i<j\leq n} (a_i b_j - a_j b_i)^2.$$

Here $a = (a_1, \ldots, a_n)$ and $b = (b_1, \ldots, b_n)$ are vectors of real numbers. Let $b = Aa$ where A is an SPD matrix. Then we have

$$\|a\|^2\|Aa\|^2 - \langle Aa, a\rangle^2 = \sum_{1\leq i<j\leq n} (a_i b_j - a_j b_i)^2$$

or equivalently for $a \neq 0$,

$$1 - \frac{\langle Aa, a \rangle^2}{\|a\|^2 \|Aa\|^2} = \frac{\sum_{1 \leq i < j \leq n} (a_i b_j - a_j b_i)^2}{\|a\|^2 \|Aa\|^2}.$$

When a is one of the two normalized first antieigenvectors $x_{\pm}^1$ of A, and let us choose x_{+}^1, namely

$$x_{+}^1 = (\lambda_n^{1/2} x_1 + \lambda_1^{1/2} x_n)(\lambda_n + \lambda_1)^{-1/2}$$

where x_1 and x_n are the extreme eigenvectors for A, then

$$Ax_{+}^1 = (\lambda_n^{1/2} \lambda_1 x_1 + \lambda_1^{1/2} \lambda_n x_n)(\lambda_n + \lambda_1)^{-1/2}$$

and $\|Ax_{+}^1\| = \lambda_n^{1/2} \lambda_1^{1/2}$. Thus the above becomes

$$1 - \cos^2 \phi(A) = \frac{\sum_{1 \leq i < j \leq n} (a_i b_j - a_j b_i)^2}{\lambda_n \lambda_1}.$$

In this way we have arrived at a new formulation of $\sin^2 \phi(A)$: the right-hand side of this expression.

By way of illustration and verification, suppose A was already diagonalized so that $x_1 = (1, 0, \ldots, 0)$ and $x_n = (0, \ldots, 0, 1)$ are the two extreme eigenvectors corresponding respectively to λ_1 and λ_n. Then

$$x_{+}^1 = \left(\left(\frac{\lambda_n}{(\lambda_1 + \lambda_n)} \right)^{1/2}, 0, \ldots, 0, \left(\frac{\lambda_1}{(\lambda_1 + \lambda_n)} \right)^{1/2} \right)$$

and

$$Ax_{+}^1 = \left(\lambda_1 \left(\frac{\lambda_n}{(\lambda_1 + \lambda_n)} \right)^{1/2}, 0, \ldots, 0, \lambda_n \left(\frac{\lambda_1}{(\lambda_1 + \lambda_n)} \right)^{1/2} \right)$$

so that the a_i and b_i in the right-hand side sum all vanish except for a_1, a_n, b_1, b_n. Hence in that simplified case this sum collapses to

$$\sum_{1 \leq i < j \leq n} (a_i b_j - a_j b_i)^2$$

$$= (a_1 b_n - a_n b_1)^2$$

$$= \left(\left(\frac{\lambda_n}{\lambda_1 + \lambda_n} \right)^{1/2} \lambda_n \left(\frac{\lambda_1}{\lambda_1 + \lambda_n} \right)^{1/2} - \left(\frac{\lambda_1}{\lambda_1 + \lambda_n} \right)^{1/2} \lambda_1 \left(\frac{\lambda_n}{\lambda_1 + \lambda_n} \right)^{1/2} \right)^2$$

$$= \left(\frac{\lambda_n - \lambda_1}{\lambda_n + \lambda_1} \right)^2 \cdot \lambda_n \lambda_1.$$

Insertion of this sum into the expression for $1 - \cos^2 \phi(A)$ above leaves you with exactly $\sin^2 \phi(A)$ on the right-hand side.

I like this result enough to be called theorem.

Theorem 9.1 (Lagrange sin $\phi(A)$). *For any $n \times n$ SPD matrix A,*

$$\sin \phi(A) = \max_{x \neq 0} \frac{\left(\sum_{1 \leq i < j \leq n} (x_i y_j - y_j x_i)^2 \right)^{1/2}}{\|x\| \|y\|}$$

where $y = Ax$.

Now we may go further and see Theorem 9.1 in the light of my program for inequality–equalities. The Lagrange identity we employed above may be seen as the inequality–equality version of the Lagrange inequality

$$\left(\sum_{i=1}^{n} |x_i|^2 \right) \left(\sum_{i=1}^{n} |y_i|^2 \right) \geq \left| \sum_{i=1}^{n} x_i y_i \right|^2$$

which holds for arbitrary complex x and y vectors. Moving the right-hand side to the left-hand side gives you the left-hand side of the Lagrange identity I used above. Thus the right-hand side of the Lagrange identity, namely, the numerator in the expression for $\sin \phi(A)$ in Theorem 9.1 above, may be regarded as that which makes up the slack. When I put in the exact antieigenvector x_+ above, all the slack terms were made equal to zero except the last term in its $i = 1$ sum. See Exercise 5.

It is a big program I am proposing here. But it can, and should, proceed in small steps. When you are using some inequality, just ask yourself: can it be improved to some meaningful inequality–equality?

9.5   Geometry

I have definitely been remiss in connecting my antieigenvalue operator trigonometry to operator geometries. Therefore there could be a rich future in doing so. What might it be?

A start was made when I recently applied my operator trigonometry directly to the geometry of quantum states. This was described in Sec. 7.6. Quantum mechanics when formulated in terms of density matrices naturally becomes an operator geometry, or perhaps stated less tersely, an operator theory with many geometrical features. I was able to augment that theory with a number of trigonometric features.

As I mentioned in [Gustafson (2011a)], there is much potential for further development of the trigonometry of quantum states to accompany the currently developing geometry of quantum states along the lines of that initial paper. One would be the development of a "second quantized" noncommutative operator trigonometry. In particular, I have already developed [Gustafson (2011c)] on a preliminary basis the essentials of such an operator trigonometry specifically in the Hilbert–Schmidt inner product $\langle A, B \rangle_F = \text{tr}(AB^*)$. I prefer to call this the Frobenius inner product. Clearly such trigonometry can be directly applied to the geometry of quantum states, and also to statistical operators acting on the density matrices. Here is a preview.

I start by recognizing the fundamental trigonometric Min-Max Theorem requirement. Because the Frobenius norm is not an induced (from vector norm) norm, I was led to begin from a notion for $\sin \phi(A)$, rather than from $\cos \phi(A)$ which had been the originating notion in 1967. Therefore let us define an operator turning angle $\theta_F(A^{1/2})$ according to

$$\sin^2 \phi_F(A^{1/2}) = \inf_{\epsilon>0} \left\| \frac{I}{\sqrt{N}} - \epsilon A^{1/2} \right\|_F^2 = \inf_{\epsilon>0} \left\{ 1 + \epsilon^2 \text{tr}(A) - \frac{2\epsilon}{\sqrt{N}} \text{tr}(A^{1/2}) \right\}.$$

Here I have formulated the trigonometry in terms of $A^{1/2}$ to render easy comparison to quantum density operators Bures distance formulations such as $D_B^2(\rho, \rho^*) = 1 + \text{tr}\,\rho - 2N^{-1/2}\,\text{tr}\,\sqrt{\rho}$. But more generally one may start directly from $\|cI - \epsilon A\|_F^2$ for various scalings c. Proceeding, one easily calculates that $f(\epsilon) \equiv \|N^{-1/2}I - \epsilon A^{1/2}\|^2$ has minimum at $\epsilon_m = \text{tr}(A^{1/2})/\sqrt{N}\,\text{tr}(A)$, and that minimum is

$$\min_{\epsilon>0} f(\epsilon) = f(\epsilon_m) = 1 - \frac{(\text{tr}\,A^{1/2})^2}{N\,\text{tr}\,A} \equiv \sin^2 \phi_F(A^{1/2}).$$

In that formulation then we have

$$\cos \phi_F(A^{1/2}) = \frac{\text{tr}\,(A^{1/2})}{(N\,\text{tr}\,A)^{1/2}}.$$

Since we know for general positive selfadjoint operators that

$$\frac{1}{N} \overset{\le}{=} \frac{(\text{tr}\,A)^2}{N\,\text{tr}\,(A^2)} \overset{\le}{=} 1,$$

such trigonometry is well defined and consistent.

Going on to formulate antieigenvectors (in this case, antieigen-matrices) is at the moment somewhat ambiguous. One can of course define them, but in the Frobenius Hilbert space, operators on operators do not have the same eigenvector (in this case, eigenmatrix) spectral properties that one is used to.

Another theme to introduce here is that of extending my operator trigonometry into new formulations specifically motivated by quantum states. It immediately stands out that because a quantum state ρ has tr $(\rho) = 1$, we could, for example, define a "mean operator angle" by

$$\cos \phi_m(A) = \frac{N \left(\prod_{i=1}^{N} \lambda_i \right)^{1/N}}{\sum_{i=1}^{N} \lambda_i} = \frac{N(\det (A))^{1/N}}{\text{tr} (A)}.$$

Then to meet the Min-Max requirement we would define an accompanying $\sin \phi_m(\rho)$ according to

$$\sin^2 \phi_m(\rho) = 1 - (N(\det \rho)^{1/N})^2.$$

In such trigonometry, only the maximally entangled centered state $\rho_* = [\text{diag}(1/N)]$ has $\cos \phi_m(\rho_*) = 1$ and hence "no turning." All other invertible density operators in the fiber bundles have $0 < \cos \phi_m(\rho) < 1$. The noninvertible density operators in the base space boundary have $\cos \phi_m(\rho) = 0$ and "90° average turning."

Stepping aside from the operator geometries of quantum mechanics, an overview into classical geometries may be seen in low dimensions. Having obtained my new formulation of $\sin \phi(A)$ in Theorem 9.1 of the preceding section, in [Gustafson (2001a)] I then briefly explored what it might mean in more conventional terms.

First, for $n = 3$, this result may be understood in terms of the well-known vector cross product:

$$\|a \times Aa\| = \|a\| \|Aa\| \sin \phi(A)$$

when $a = x^1_{\pm}$ is an antieigenvector of A. Thus in three dimensions we have a way to see $\sin \phi(A)$ with exactly the same geometrical content as that for $\cos \phi(A) : \langle Aa, a \rangle = \|a\| \|Aa\| \cos \phi(A)$, for $a = x^1_{\pm}$ an antieigenvector of A.

Second, as is well-known, vector cross products only exist for the special vector space dimensions $n = 1, 3$, and 7. However, we really do not need to be in a cross product dimension. For arbitrary dimensions one may go back to the old theory of "compounds" in the theory of minor expansions of determinants.

Third, for $n = 2$ we may obtain a "symplectic" representation for $\sin \phi(A)$. For, from Theorem 9.1 one may verify that

$$\sin \phi(A) = \max_{x \neq 0} \frac{\langle Ax, J^*x \rangle}{\|Ax\| \|J^*x\|}$$

where $J = \begin{bmatrix} 0 & 1 \\ -1 & 0 \end{bmatrix}$. For $n = 3$ in a similar way one can express the sum in Theorem 9.1 in an analogous way. That is, consider the sum $\sum (a_i b_j - a_j b_i)^2$, $1 \leq i < j \leq 3$, $x = (a_1, a_2, a_3)$, $Ax = (b_1, b_2, b_3)$. Then

$$\sum_{1 \leq i < j \leq 3} (a_i b_j - a_j b_i)^2 = \langle Ax, J_1 x \rangle^2 + \langle Ax, J_2 x \rangle^2 + \langle Ax, J_3 x \rangle^2$$

where

$$J_1 = \begin{bmatrix} 0 & -1 & 0 \\ 1 & 0 & 0 \\ 0 & 0 & 0 \end{bmatrix}, \quad J_2 = \begin{bmatrix} 0 & 0 & -1 \\ 0 & 0 & 0 \\ 1 & 0 & 0 \end{bmatrix}, \quad J_3 = \begin{bmatrix} 0 & 0 & 0 \\ 0 & 0 & -1 \\ 0 & 1 & 0 \end{bmatrix}.$$

Since symplectic theory holds for even dimensions one could work out similar representations $n = 4, 6, \ldots$.

Fourth, similar formulations hold for complex vector spaces. Quaternions, Clifford algebras and spin theories are waiting to be investigated.

Some other incomplete investigations discussed in the paper [Gustafson (2011c)] will be the following.

There is Fricke's theory of characters for subgroups of the special linear group SL(2, K) over a commutative ring K, see e.g., [Magnus (1974)]. For 2×2 matrices A_i, $i = 1, \ldots, n$, all with determinant 1, then for any matrix M in the group that the A_i's generate, tr M is a polynomial with coefficients determined by the traces of the A_i's. For example,

$$\text{tr} (A_1 A_2 A_1^{-1} A_2^{-1}) = x^2 + y^2 + z^2 - xyz + 2$$

where $x = \text{tr } A_1$, $y = \text{tr } A_2$, $z = \text{tr } (A_1 A_2)$. This expression is so close to the Gram matrix expression (∗∗∗)

$$1 + 2a_1 a_2 a_3 - a_1^2 - a_2^2 - a_3^2 \geq 0$$

of Sec. 3.5 that there must be valuable connections between the Fricke–Klein group characters and the operator trigonometry.

In this spirit, we also may look at shape invariants of triangles. Following [Brehm (1990)], consider the shape invariant $\sigma(x,y,z) = \mathrm{Re}\,\langle x,y\rangle\langle y,z\rangle\langle z,x\rangle$. Then Brehm's equation (2), but stated now in terms of my trigonometric context of Sec. 3.5, becomes

$$\cos^2 a + \cos^2 b + \cos^2 c - 1 \overset{\leq}{=} 2|\sigma| \overset{\leq}{=} 2\cos a \cos b \cos c.$$

Pushing further, that triangular theory of shape invariants may be viewed from my Gram determinant inequalities, from which we know just from $|G| \overset{\geq}{=} 0$ that

$$2a_1 a_2 a_3 \overset{\geq}{=} a_1^2 + a_2^2 + a_3^2 - 1.$$

Thus I may even interpret the shape invariant $2|\sigma|$ as a slack variable which must interpolate between the two sides of the operator trigonometric triangle inequality.

One should go further with this, e.g., to SO(2) and SU(3), for example, and also even back to classical spherical trigonometry, and then off into general homogeneous symmetric spaces. For example, in [Ortega and Santander (2002)] there is a study of the trigonometry of the Cayley–Klein–Dickson family of spaces including $\mathbb{C}P^N$ and $\mathbb{C}H^N$. Lo and behold, in (6.49) there I find the Gram matrix expansion of our operator trigonometry, but now in the form

$$\Delta_g = 1 - C^2(x_1) - C^2(x_2) - C^2(x_3) + 2C(x_1)C(x_2)C(x_3)C(\Omega),$$

as a key entity in the classification of CKD spaces from the viewpoint of Cartan sector quantities.

One should never hesitate to bring things back to Earth and reground yourself on *terra firma*. Although we were led to our matrix Gram determinant expression (∗∗∗) in Sec. 3.5 in order to prove our triangle inequality of Theorem 3.3, in view of the potential connections to more general shape invariants in group theory, where, from that point of view, have we really come from? Remember the admonition here is to be as concretely concrete (excuse the double emphasis) as possible. So I looked back into my old elementary solid geometry book and found the following. Given a tetrahedron with one vertex at the origin and the other three vertices at vectors $a, b,$ and c in three-space, then its volume is given by

$$V = \frac{|a|\|b\|\|c\|}{6}[1 + 2\cos\alpha\cos\beta\cos\gamma - \cos^2\alpha - \cos^2\beta - \cos^2\gamma]^{1/2}$$

where the α, β, γ are the subtended angles at the origin.

Originally, I always thought of determinants of Gram matrices for expressing volumes of rectangles and their skewed parallelepipeds. But a tetrahedron is a simplex in three-dimensional space, and those shapes have distinct advantages over rectangles and parallelepipeds in the finite element approximation theory of partial differential equations. Rectangles (called "bricks" in the FEM literature) do not have the nice combinational structure of tetrahedral meshes, nor do they approximate boundaries as well. See for example [Gustafson (1983b)] for some of these dimensional features of finite element meshes. So I would expect interesting local operator trigonometries to be found within those simplicial meshes, on which, after all, the partial differential operator is represented by loosely interconnected matrix operators.

Even more interesting would be to use this tetrahedral bridge into new applications of the operator trigonometry to, for example, chemistry or vision, wherein the roles of tetrahedral concepts is already established.

9.6 Applications

I can summarize this section in one sentence. I want more applications of my antieigenvalue theory. That was the main goal of writing this monograph. I will be equally happy should those new applications come from anywhere: pure mathematics, applied mathematics, computational mathematics, physics, engineering, economics, psychology.... One outcome of the writing of this book which might make it easier for nonspecialists to engage my theory is that instead of my usual terminology of matrix turning angles, they are better seen, felt, as matrix twisting angles.

An interesting recent application from the realm of engineering was the use of my antieigenvalue theory by Cuntoor and Chellappa (2006) to detect key frames in motion analysis. Both the extent of change as given by the antieigenvalues, and the location of change as given by the key frames, are useful for recognition.

In a lecture presented at the IWMS 2010 Conference at the Shanghai Finance University where I gave the survey address [Gustafson (2011b)], H. Yanai of Japan announced his talk as "On the inequality $1 + 2abc - a^2 - b^2 - c^2 \geqq 0$." Of course I was surprised and interested and attended that session, where the official title changed to that of his preprint [Yanai,

Puntanen, Ito and Ishii (2008)]: Some extensions on the ranges of a correlation matrix and its inverse matrix. Therein you will find a number of implications of the positivity of the Gram matrix, which we used originally in 1969 to obtain our triangle inequality of Sec. 3.5, toward statistical partial correlation, multiple correlation, and canonical correlation coefficients. The a, b, c of the inequality are the partial correlations between three random variables X, Y, and Z. Such investigations within the statistics literature go back to [Olkin (1981)] and further.

Then of course there will be plain-vanilla applications, just for their own sakes, to items from pure mathematics which happen to meet the eye. These might just be idle exercises, or they may lead to somewhere interesting. An example has been placed into Exercise 6 to end this chapter.

Commentary

As I stated in the *Perspective* to this last chapter, the chapter would attempt to be indicative of other and future directions for the antieigenvalue analysis and its associated noncommutative operator trigonometry, and above all . . . be incomplete. I hope so. One does not want a closed future.

So why say more? Let the book set sail.

9.7   Exercises

1. Verify that in the re-inner-product space mentioned in Sec. 9.1, namely,

$$\langle x, y \rangle_B = \langle Bx, y \rangle$$

where $B = (SS^*)^{-1}$, that B is SPD in the original inner product space. Then show that in the new inner product space, the dual operator for A is

$$A' = B^{-1}A^*B$$

where A^* was the dual in the original inner product.

2. In an interesting microcosmos model formulated by physics Nobel Laureate Gerard 't Hooft (2002), he proposes that information loss at the Planck scale may be an essential ingredient in deterministic hidden

variables. To fix ideas, he proposes an evolution matrix

$$U = \begin{bmatrix} 0 & 0 & 1 & 0 \\ 1 & 0 & 0 & 1 \\ 0 & 1 & 0 & 0 \\ 0 & 0 & 0 & 0 \end{bmatrix}.$$

This may be thought of as the directed adjacency matrix for a mini-universe with information loss. Show that U is a normalizable operator.

3. Further to Sec. 9.3, verify that the optimizing Kantorovich–Wielandt vectors are equally weighted, viz., they are

$$x = x_1 + x_n, \quad y = x_1 - x_n.$$

4. Verify that

$$x_{\pm} = \left(\frac{\lambda_1^{1/2}}{\lambda_1^{1/2} + \lambda_n^{1/2}} \right)^{1/2} x_1 \pm \left(\frac{\lambda_n^{1/2}}{\lambda_1^{1/2} + \lambda_n^{1/2}} \right)^{1/2} x_n$$

do indeed optimize the Shisha–Mond inequality. Play with trying to reconcile how these, the antieigenvectors of the operator $A^{-1/2}$, show up to optimize this inequality. Why not, say, $A^{1/2}$?

5. Write out the details of the collapse of the slack variables in the Theorem 9.1 formulation of $\sin \phi(A)$ when one puts in the antieigenvector x_+.

6. Here are three final easy exercises of the "just for fun" genre.

   (a) In his recent paper Coudene (2006) points out that there are almost no pictures of uniformly hyperbolic attractors of the Anisov type from the theory of dynamical systems. Some are provided in that paper. The first example is a perturbation (due to S. Smale) of a hyperbolic automorphism of the two-torus T^2. The starting system involves the matrix $A = \begin{bmatrix} 2 & 1 \\ 1 & 1 \end{bmatrix}$, written in its singular value decomposition (i.e., spectral representation) as

$$\begin{bmatrix} 2 & 1 \\ 1 & 1 \end{bmatrix} = \frac{1}{(1+\lambda^2)^{1/2}} \begin{bmatrix} \lambda & -1 \\ 1 & \lambda \end{bmatrix} \begin{bmatrix} \lambda^2 & 0 \\ 0 & \lambda^{-2} \end{bmatrix} \frac{1}{(1+\lambda^2)} \begin{bmatrix} \lambda & 1 \\ -1 & \lambda \end{bmatrix}.$$

A has eigenvalues λ^2 and λ^{-2}, where λ is the golden mean

$$\lambda = \frac{1+\sqrt{5}}{2} \cong 1.618.$$

The corresponding eigenvectors, respectively, are

$$x_2 = \frac{1}{(1+\lambda^2)^{1/2}} \begin{bmatrix} \lambda \\ 1 \end{bmatrix}, \quad x_1 = \frac{1}{(1+\lambda^2)^{1/2}} \begin{bmatrix} -1 \\ \lambda \end{bmatrix}.$$

Work out some operator trigonometry for this example.

(b) For an arbitrary SPD $n \times n$ matrix A, show that the antieigenvector set $\{x_\pm\}$ becomes an orthonormal basis in the A-inner product.

(c) Combine (a) and (b) above to show that the $\{x_\pm\}$ antieigenvectors in the golden mean A above are indeed orthogonal in the A-inner product. Also confirm that in the original inner product, $\langle x_+, x_- \rangle = \sin \phi(A)$ in accordance with Lemma 5.1 (see Appendix B, Exercise 6 of Chapter 5). Recall that therein we used the convention

$$x_\pm = \left(\frac{\lambda_{max}}{\lambda_{max} + \lambda_{min}} \right)^{1/2} x_{min} \pm \left(\frac{\lambda_{min}}{\lambda_{max} + \lambda_{min}} \right)^{1/2} x_{max},$$

for the two antieigenvectors, i.e., putting the $\pm$ on the second term.

Linear Algebra

Linear Algebra is a huge subject nowadays and also touches a gamut of applications, ranging from representation theory in pure group theory to the most advanced computational science. Here in Appendices A.1 and A.2 I very briefly bring to mind the contexts of Matrix Analysis and Operator Theory as I see them, thereby helping to explain the flavors of this book.

A.1   Matrix Analysis

A typical first semester undergraduate Linear Algebra course should start with matrix-vector multiplication and teach the Gauss reduction procedure to solve linear systems. One should then introduce the axioms of vector spaces and of algebras and show that the $n \times n$ matrices are both. Then the Fundamental Theorem of Linear Algebra, that $R(A) = N(A^*)^\perp$, should be established. Next, eigenvalue–eigenvector problems should be posed and solved for both symmetric and nonsymmetric matrices A. Finally, the singular value decomposition, and as many of its ramifications as time permits, should be given ample study.

The next (graduate) course should immediately emphasize how to change bases, while at the same time stressing the need to pick a good one to calculate anything useful. Schur's upper-triangularization theorem and the Jordan canonical form should be proved. I might next go to computational linear solvers, but others teaching such a course could go to other topics, for example search engines. They are too many to be listed here.

What are good textbooks? There are so many good ones out there that when publishers approached me to write my own, I decided not to. What have I found myself choosing as textbook? I have chosen the undergraduate text by D. Lay, *Linear Algebra and its Applications*, and the graduate text by

R. Horn and C. Johnson, *Matrix Analysis*. Both of these excellent books have proven themselves to wide audiences.

A.2   Operator Theory

I was raised on S. Goldberg's little classic, *Unbounded Linear Operators*, although before that I had been exposed to the classic *Functional Analysis* by F. Riesz and B. S. Nagy. Of course I like Kato's *Perturbation Theory for Linear Operators*. When I taught a course in functional analysis many years ago I found myself choosing Yosida's *Functional Analysis*. It would appear that I prefer the older classics to any of the newer texts I have chanced upon. However, I am sure there are some good new ones. If I were to teach such course again, I would probably not be able to resist P. Lax's *Functional Analysis*, just because of its interesting content.

How do I see operator theory? Because of my background in partial differential equations and its applications, I see unbounded operators in a Hilbert space as the place to start. Then their resolvent operators $(\lambda I - A)^{-1}$ take you to bounded operators. Then if you actually need to compute things useful in real applications, you will specialize further to finite-dimensional matrices. Quickly you will find yourself immersed in linear solvers for very large but finite matrix problems. You could spend the rest of your life working within the enterprise of linear solvers. I am astonished at some of my colleagues in algebra who do not even know of the huge field of computational linear algebra. At least I have worked with Galois fields for error correction, and with Lie algebras for differential equations. Perhaps Linear Algebra is now the most important topic in Algebra.

Hints and Answers to Exercises

Chapter 1.

1. Schwarz's inequality is perhaps the most important inequality in mathematics, hence, in all the world. If you just care about real Hilbert spaces, I give the following very short proof in my book *Partial Differential equations* [Gustafson (1999f)]. Let $Q(t) = \langle u + tv, u + tv \rangle = \|u\|^2 + t(2\langle u, v \rangle) + t^2\|v\|^2$. As $Q(t)$ is never negative, Schwarz's inequality is just the nonpositivity of $Q(t)$'s discriminant. You will find other inequalities we assume in this book, in that PDE book, and will notice from Chapter 1 that I prefer the notation $\langle x, y \rangle$ for inner product.

2. There are several ways to approach this, notably, by functional minimization, or just using ordinary calculus. Let us do the latter here. Consider $R(x)$ denote the Rayleigh quotient for the matrix $A = [a_{ij}]$. Let $x_1 = [x_i^1]$ be any lowest eigenvector. Then $R(x)$ is a continuous, even differentiable, function of all of the coordinates x_i. $R(x)$ achieves its minimum at x_1. Hence, for $i = 1, 2, \ldots, n$, we have necessarily the system

$$0 = \frac{\partial R(x)}{\partial x_i} = \frac{\partial}{\partial x_i}\left[\frac{\sum_{i,j=1}^n a_{kj}a_k x_j}{\sum_{i=1}^n x_i^2}\right]_{x_i=x_i^1} = 2\left[\frac{\sum_{i=1}^n a_{ij}x_j^1 - x_i^1 R(x_1)}{\sum_{k=1}^n (x_k^1)^2}\right]$$

$$= Ax_1 - \lambda x_1.$$

You may do the higher antieigenvalues the same way on the reduced matrix on the reduced spaces.

3. Everyone is familiar with the spectral representation

$$A = \sum_{i=1}^{n} \lambda_i x_i x_i^* = \sum_{i=1}^{n} \lambda_i P_i, \quad P_i = P_i^2 = P_i^*,$$

for selfadjoint matrices. But many do not keep cognizance of the more general theorem expressing A in terms of rank-one outer products

$$A = \sum_{k=1}^{k} x_i y_i^*$$

for arbitrary matrices of rank k. I like to get my version of this just by a judicious Gauss elimination procedure: Let $EA = E_k E_{k-1} \cdots E_1 A = Y$ be a reduction of A to an upper echelon form Y, so that $A = XY$ where $X = E_1^{-1} \cdots E_{k-1}^{-1} E_k^{-1}$. Then write the matrix product XY as outer-products. Thus $A = XY = X_1 Y_1^* + \cdots + X_k Y_k^*$ where the X_i are the columns of X and the Y_i the rows of Y. Then I like to go further and normalize to weighted projections

$$A = \sum_{i=1}^{k} (y_i^* x_i) \frac{x_i y_i^*}{y_i^* x_i} = \sum_{i=1}^{k} \langle x_i, y_i \rangle P_x^{y^\perp}.$$

Here each P_i is the oblique projector onto the span of x, projecting from the direction $y^\perp$. That is the reason for my notation $P_x^{y^\perp}$.

Here is an example. Let

$$x = \begin{bmatrix} 1 \\ 0 \end{bmatrix}, \quad y = \begin{bmatrix} 1 \\ 1 \end{bmatrix}, \quad w = \begin{bmatrix} 1 \\ 1/2 \end{bmatrix}.$$

Then for any x, y, w, we have the projector implemented in either of two ways:

$$P_x^{y^\perp} w = \frac{xy^*}{y^* x}(w) = \frac{(y^* w)}{(y^* x)} x.$$

For the specific vectors $x, y,$ and w above, we have

$$P_x^{y^\perp}(w) = \begin{bmatrix} 1 & 1 \\ 0 & 0 \end{bmatrix} \begin{bmatrix} 1 \\ 1/2 \end{bmatrix} = \begin{bmatrix} 3/2 \\ 0 \end{bmatrix}$$

and

$$P_x^{y^\perp}(w) = \left([1 \ 1] \begin{bmatrix} 1 \\ 1/2 \end{bmatrix} \right) \begin{bmatrix} 1 \\ 0 \end{bmatrix} = \begin{bmatrix} 3/2 \\ 0 \end{bmatrix}.$$

It is nice to draw the picture of this simple example.

One need not use my Gauss elimination recipe above, and indeed the possibilities for such oblique projections are rich. How you choose to do will depend upon your needs. For example, one may just spectrally project onto the columns of a matrix A, e.g.,

$$A = \begin{bmatrix} 1 & 2 \\ 3 & 1 \end{bmatrix} = \begin{bmatrix} 1 & 0 \\ 3 & 0 \end{bmatrix} + \begin{bmatrix} 0 & 2 \\ 0 & 1 \end{bmatrix} = \begin{bmatrix} 1 \\ 3 \end{bmatrix} [1 \ 0] + \begin{bmatrix} 2 \\ 1 \end{bmatrix} [0 \ 1].$$

Orthogonal projectors P have long ruled for many good reasons. We like to think in terms of the natural basis $\{e_i\}$. Generally they provide better round-off accuracy because they have $\|P\| = 1$. But as we move more and more into general matrices, oblique projectors will more often naturally occur in their own right.

4. My favorite book on unbounded operator theory is [Kato (1976)].
5. This is even true in infinite dimensions. See [Halmos (1967)].
6. This is an infinite-time problem, so just follow your interests.

Chapter 2.

1. One may employ Green's identity

$$0 \overset{\leq}{=} \int_\Omega (\operatorname{grad} u)^2 = \oint_{\partial\Omega} u \frac{\partial u}{\partial n} - \int_\Omega u \Delta u,$$

that holds for smooth functions u. Then you must go to the full domain $W^{2,2}(\Omega)$ of the operator A.

On bounded domains Ω one can show that the operator $-\Delta$ is strongly accretive. The Rayleigh quotient for functions u that vanish on the boundary $\partial\Omega$ is nonnegative, i.e.,

$$\frac{\langle -\Delta u, u \rangle}{\langle u, u \rangle} = \frac{\int_\Omega (\operatorname{grad} u)^2}{\int_\Omega u^2} \overset{\geq}{=} 0.$$

Suppose it is zero for some u in the domain of the operator. Then the gradient $\operatorname{grad} u$ is zero at each point in Ω, hence u is constant over Ω, and hence by continuity to the boundary, u is zero on the closed domain $\bar{\Omega}$, a contradiction. Stated another way, the operator is selfadjoint with positive lowest eigenvalue λ_1. We have glossed over the need to take this argument from a smooth dense subspace of functions u to the whole domain of the operator.

2. See any ordinary differential equations textbook.
3. For an overly brief account, see the Epilogue in my PDE book [Gustafson (1999f)].
4. See my book [Gustafson (1999f)]. For a much more comprehensive treatment see [Kato (1976)].
5. For A an unbounded operator, the statement that A commutes with bounded everywhere defined B means the operator product inclusion

$$BA \subset AB.$$

The domain of the left-hand side is $D(A)$. Normally the domain of AB depends on the meshing of the range $R(B)$ with the domain $D(A)$. So a commuting assumption is even stronger than its face value would indicate. In particular, it means that $D(AB)$ is at least as large as $D(A)$.

A possesses positive square root $A^{1/2}$ such that

$$BA^{1/2} \subset A^{1/2}B.$$

Thus $\langle BAx, x \rangle = \langle BA^{1/2}x, A^{1/2}x \rangle$ on $D(BA) = D(A) \subset D(A^{1/2})$. Hence $\mathrm{Re}\,\langle BAx, x \rangle \overset{\geq}{=} 0$ due to the accretivity of B.

As to the second assertion, first we establish that BA is symmetric. For bounded operators B, one always has the adjoint inclusion relation $(BA)^* \supset A^*B^*$. For the case at hand, from $BA \subset AB$ we have

$$(BA)^* = AB^* \supset (AB)^* \supset B^*A.$$

Thus in particular, A also commutes with B^*. If we now impose the stronger hypothesis that B is selfadjoint and strongly positive, then the inclusion above becomes

$$(BA)^* = AB \supset (AB)^* \supset BA$$

and therefore BA is a symmetric operator.

By the first part of the problem, BA is already accretive. If BA is moreover selfadjoint, then its accretivity turns into positivity (nonnegativity to be more precise). That turns into strong positivity from the strong positivity of B and that of A, hence of $A^{1/2}$: for x in $D(A)$ we have

$$\langle BAx, x \rangle = \langle BA^{1/2}x, A^{1/2}x \rangle \overset{\geq}{=} m_B \langle A^{1/2}x, A^{1/2}x \rangle$$
$$= m_B \langle Ax, x \rangle \overset{\geq}{=} m_B m_A \langle x, x \rangle.$$

I have used the fact that when we say a selfadjoint operator is positive, that implicitly implies that A is strongly positive. That is the case when

A is bounded, but probably for unbounded A one should more carefully just make the assumption of strong positivity.

In any case, it remains to show that BA is selfadjoint. For this we may use the Rellich–Kato additive perturbation theorem. For small positive ϵ, write as I did in the multiplicative perturbation proofs

$$\epsilon BA = (\epsilon B - I)A + A.$$

Since $\|\epsilon B - I\| = b < 1$, A is selfadjoint and the perturbation $(\epsilon B - I)A$ is symmetric, by the Rellich–Kato Theorem the right-hand side is selfadjoint. No doubt there are other proofs but I wanted an exercise on adjointing and commuting of unbounded operators and a use of the Rellich–Kato Theorem.

6. See the paper [Gustafson (2007a)]. As I recall at this writing, I have not gotten around to the next step, namely, of developing the operator trigonometry which might result by utilizing this graph norm-to-norm unbounded closed operators.

Chapter 3.

1. There are many proofs of this theorem in the literature. I like my own, [Gustafson (1970a)], because it uses the same mindset as that which I used earlier in proving my Min-Max Theorem [Gustafson (1968d)].

2. The corrected expression on p. 163 of [Gustafson (1968c)] is

$$\|\epsilon B + I\| \lesseqgtr \|I - \epsilon^2 B^2\| \cdot \|(\epsilon B - I)^{-1}\| \lesseqgtr (1 + \epsilon^2 \|B\|^2) \cdot (1 - \epsilon \theta(B))^{-1}.$$

3. Let $\mathbb{R}^2$ have a "skew" unit ball. For a concrete example, draw the unit ball as follows. Starting at the point $(x_1, x_2) = (-1, -1)$, come horizontally to the right to $(x_1, x_2) = (1/2, -1)$, then go up to the right at 60° to the point $(x_1, x_2) = (1, 0)$, go straight up to $(x_1, x_2) = (1, 1)$, turn horizontally left all the way back to $(x_1, x_2) = (-1/2, 1)$, descend at 60° to the x_1 axis at $(x_1, x_2) = (-1, 0)$, then drop vertically down to the first point $(x_1, x_2) = (-1, -1)$.

Now let B be the 2×2 matrix $B = \begin{bmatrix} 1/2 & 0 \\ 0 & 1 \end{bmatrix}$. Then letting ϵ trace through the values $\epsilon = 1, 5/4, 3/2$, one may verify that the $\|\epsilon B - I\|$ curve is exactly the line $1 - \epsilon/2$ for $0 \lesseqgtr \epsilon \lesseqgtr 1$, then the flat of value $1/2$ for $1 \lesseqgtr \epsilon \lesseqgtr 3/2$, then the increasing line $-1 + \epsilon$ for $\epsilon \gtreqless 3/2$.

4. From $F(\lambda) = (2 + \lambda^2)/(\lambda^4 + 5\lambda^2 + 4)^{1/2}$ we obtain

$$F'(\lambda) = \frac{\lambda^3 - 2\lambda}{(\lambda^4 + 5\lambda^2 + 4)^{3/2}}$$

and

$$F''(\lambda) = \frac{(\lambda^4 + 5\lambda^2 + 4)(3\lambda^2 - 2) - 3(\lambda^3 - 2\lambda)(4\lambda^3 + 10\lambda)/2}{(\lambda^4 + 5\lambda^2 + 4)^{5/2}}.$$

$F'(\lambda) = 0$ at $\lambda = 0$ or $\lambda = \pm\sqrt{2}$. The former comes from the largest eigenvector which creates the $F(\lambda) = 1$ maximum at $\lambda = 0$. The latter come from the two antieigenvectors and create the minima as shown in Fig. 3.2. To obtain the inflection points we set the numerator of $F''(\lambda) = 0$ from which we obtain the sixth-order polynomial equation

$$3\lambda^6 - 10\lambda^4 - 32\lambda^2 + 8 = 0.$$

I could not quickly factor this so I went to the computer to find the approximate positive roots $\lambda^2 = 0.23408$ and 5.26358, whose square roots provide the inflection points for Fig. 3.2.

As $\lambda \to \infty$ we see that

$$F(\lambda) = \frac{1 + 2/\lambda^2}{1 + 5/\lambda^2 + 4/\lambda^4} \to 1$$

and

$$F'(\lambda) = \frac{1 - 2/\lambda^2}{(\lambda^2 + 5 + 4/\lambda^2)^{3/2}} \to 0.$$

5. You should start from scratch by putting your $x_{\pm}^{jk}$ directly into the Euler equation. However, starting from our expressions in proving the converse, you have immediately that

$$|c_1|^2\lambda_1^2 + |c_2|^2\lambda_2^2 = \lambda_2\lambda_1,$$

$$|c_1|^2\lambda_1 + |c_2|^2\lambda_2 = \frac{2\lambda_2\lambda_1}{\lambda_1 + \lambda_2},$$

and hence the coefficient of x_1 becomes simply

$$\frac{\lambda_1}{\lambda_2} - \frac{2(\lambda_1 + \lambda_2)}{2\lambda_2} + 1 = 0.$$

Similarly for the other coefficient.

6. I must confess that I have not worked out the particulars. And there may be interesting unforseen consequences in considering three-component and k-component generalizations of my antieigenvectors. However, keep in mind that physical intuition says that if you try to

turn a three-component vector x with A, the "middle" component will get in the way and you will therefore turn more by dropping the middle component from the trial vector. So one needs to look for trigonometrical features other than just maximum turning dynamics.

Chapter 4.

1. D. M. Young is the older and very famous computational linear algebraist from Texas and Ph.D. student of G. Birkhoff at Harvard. His book which really put iterative methods on the map and which contributed SOR to our mathematics is D. M. Young, *Iterative Solutions of Large Linear Systems*, Academic Press, 1971.

   D. P. Young was my Ph.D. student (1979) at Colorado. He was so good that Boeing snapped him up immediately, and to my knowledge David is still there and has been a key design engineer for CFD projects such as the Space Shuttle, Stealth, and Boeing 777.

   I first met the senior D. M. Young at the Nijmegen conference in 1996. He humorously castigated me for the resulting confusion caused by my bringing another D. Young into the computational linear algebra literature!

2.

$$\sin^2 \phi(A^{1/2}) = \left[\frac{\lambda_n^{1/2} - \lambda_1^{1/2}}{\lambda_n^{1/2} + \lambda_1^{1/2}} \right]^2 = \frac{\lambda_n + \lambda_1 - 2\lambda_n^{1/2}\lambda_1^{1/2}}{\lambda_n + \lambda_1 + 2\lambda_n^{1/2}\lambda_1^{1/2}}.$$

Thus

$$\sin \phi(A^{1/2}) = \left[\frac{1 - \cos \phi(A)}{1 + \cos \phi(A)} \right]^{1/2} = \frac{\sin \phi(A)}{1 + \cos \phi(A)}.$$

3. See [Gustafson (1998g)] for details. Block ordering A produced a better angular route to the solution.

4. The eigenvalues of A_h are

$$\lambda_{ij} = 4h^{-2} \left[\sin^2 \left(\frac{i\pi h}{2} \right) + \sin^2 \left(\frac{j\pi h}{2} \right) \right], \quad 1 \stackrel{\le}{=} i, \quad j \stackrel{\le}{=} N - 1.$$

The corresponding i,j-th eigenvectors e^{ij} are the $(N-1)^2$ dimension vectors with components

$$(e^{ij})_{k\ell} = 2h \sin (i\pi hk) \sin (j\pi h\ell),$$

with $1 \stackrel{\le}{=} k, \ell \stackrel{\le}{=} N - 1$ taken in natural order.

Thus the smallest and largest eigenvalues are

$$\lambda_1 = \lambda_{11} = 8h^{-2} \sin^2\left(\frac{\pi h}{2}\right),$$

$$\lambda_n = \lambda_{N-1,N-1} = 8h^{-2} \sin^2\left(\frac{(N-1)\pi h}{2}\right)$$

$$= 8h^{-2} \cos^2\left(\frac{\pi h}{2}\right).$$

For the coefficients of the antieigenvectors we need

$$\left(\frac{\lambda_n}{\lambda_1 + \lambda_n}\right)^{1/2} = \frac{\cos(\pi h/2)}{(\sin^2(\pi h/2) + \cos^2(\pi h/2))^{1/2}} = \cos\left(\frac{\pi h}{2}\right),$$

$$\left(\frac{\lambda_1}{\lambda_1 + \lambda_n}\right)^{1/2} = \frac{\sin(\pi h/2)}{(\sin^2(\pi h/2) + \cos^2(\pi h/2))^{1/2}} = \sin\left(\frac{\pi h}{2}\right).$$

Thus

$$x_{\pm} = \pm \cos\left(\frac{\pi h}{2}\right) e_{11} + \sin\left(\frac{\pi h}{2}\right) e_{N-1,N-1}.$$

For the 4×4 coarse mesh example with $h = 1/3$ one has

$$\lambda_{11} = \frac{8}{9} \sin^2\left(\frac{\pi}{6}\right) = \frac{2}{9},$$

$$\lambda_{22} = \frac{8}{9} \cos^2\left(\frac{\pi}{6}\right) = \frac{6}{9},$$

$$e_{11} = \frac{2}{3}\begin{bmatrix} \sin\left(\frac{\pi}{3}\right)\sin\left(\frac{\pi}{3}\right) \\ \sin\left(\frac{2\pi}{3}\right)\sin\left(\frac{\pi}{3}\right) \\ \sin\left(\frac{\pi}{3}\right)\sin\left(\frac{2\pi}{3}\right) \\ \sin\left(\frac{2\pi}{3}\right)\sin\left(\frac{2\pi}{3}\right) \end{bmatrix} = \frac{1}{2}\begin{bmatrix} 1 \\ 1 \\ 1 \\ 1 \end{bmatrix},$$

$$e_{22} = \frac{2}{3}\begin{bmatrix} \sin\left(\frac{2\pi}{3}\right)\sin\left(\frac{2\pi}{3}\right) \\ \sin\left(\frac{4\pi}{3}\right)\sin\left(\frac{2\pi}{3}\right) \\ \sin\left(\frac{2\pi}{3}\right)\sin\left(\frac{4\pi}{3}\right) \\ \sin\left(\frac{4\pi}{3}\right)\sin\left(\frac{4\pi}{3}\right) \end{bmatrix} = \frac{1}{2}\begin{bmatrix} 1 \\ -1 \\ -1 \\ 1 \end{bmatrix}.$$

From these we obtain

$$x_{\pm} = \pm \left(\frac{6}{2+6}\right)^{1/2} e_{11} + \left(\frac{2}{2+6}\right)^{1/2} e_{22} = \pm \left(\frac{\sqrt{3}}{2}\right) e_{11} + \frac{1}{2} e_{22},$$

which more specifically are

$$x_{+} = \frac{1}{4} \begin{bmatrix} \sqrt{3}+1 \\ \sqrt{3}-1 \\ \sqrt{3}-1 \\ \sqrt{3}+1 \end{bmatrix} \cong \begin{bmatrix} 0.68301 \\ 0.18301 \\ 0.18301 \\ 0.68301 \end{bmatrix}$$

and

$$x_{-} = \frac{1}{4} \begin{bmatrix} -\sqrt{3}+1 \\ -\sqrt{3}-1 \\ -\sqrt{3}-1 \\ -\sqrt{3}+1 \end{bmatrix} \cong \begin{bmatrix} -0.18301 \\ -0.68301 \\ -0.68301 \\ -0.18301 \end{bmatrix}.$$

5. Starting from the result of Chatelin and Gratton (2000), we have

$$C^2(H) = \frac{2(1 + \kappa^2(A))}{(1 + \kappa(A))^2} = \frac{2(1 + \sigma_1^2(A)/\sigma_n^2(A))}{(1 + \sigma_1(A)/\sigma_n(A))^2}$$

$$= \frac{2(\lambda_1^2(H) + \lambda_n^2(H))}{(\lambda_1(H) + \lambda_n(H))^2} = 1 + \left(\frac{\lambda_n(H) - \lambda_1(H)}{\lambda_n(H) + \lambda_1(H)}\right)^2.$$

6. Hint: See [Gustafson (2004a)]. There I characterized the positive cone $\{C\}$ by a Jordan canonical form analysis in which I wanted AB in its upper Jordan form $\widetilde{AB}$. To get there for the 2×2 example of the problem at hand, note that

$$BA = \begin{bmatrix} 1 & 2/7 \\ 1/2 & 1 \end{bmatrix}, \quad AB = \begin{bmatrix} 1 & 1/2 \\ 2/7 & 1 \end{bmatrix}$$

and $\sigma(AB) = \sigma(BA)$ are the eigenvalues

$$\lambda_1 = 1 - \frac{1}{7}, \quad \lambda_2 = 1 + \frac{1}{7}.$$

Seeking all similarity matrices M such that $(AB)M = M(\widetilde{AB})$ means we want

$$\begin{bmatrix} 1 & 1/2 \\ 2/7 & 1 \end{bmatrix} \begin{bmatrix} m_{11} & m_{12} \\ m_{21} & m_{22} \end{bmatrix} = \begin{bmatrix} m_{11} & m_{12} \\ m_{21} & v_{22} \end{bmatrix} \begin{bmatrix} 1 - \sqrt{7}/7 & 0 \\ 0 & 1 + \sqrt{7}/7 \end{bmatrix}.$$

One eventually arrives at the expression

$$C = \begin{bmatrix} m_{11}^2 \lambda_1 + m_{22}^2 \lambda_2(7/4) & m_{11}^2 \lambda_1(-2/\sqrt{7}) + m_{22}^2 \lambda_2(\sqrt{7}/2) \\ m_{11}^2 \lambda_1(-2/\sqrt{7}) + m_{22}^2 \lambda_2(\sqrt{7}/2) & m_{11}^2 \lambda_1(4/7) + m_{22} \lambda_2 \end{bmatrix}.$$

See [Gustafson (2004a)] for details.

Chapter 5.

1. Recall that our Time-operator for any wavelet is

$$Tf = \sum_{m \in \mathbb{Z}} \sum_{n \in \mathbb{Z}} m \langle \psi_{mn}, f \rangle \psi_{mn}(x).$$

The Haar wavelet $\psi_{1,1}$ oscillates once with amplitude $2^{-3/2}$ on its support $[2, 4]$ and the wavelet $\psi_{3,0}$ oscillates once with amplitude $2^{-3/2}$ on its support $[0, 8)$. Our new example of a wavelet twisting vector will be my antieigenvector (I just do the x_+ positive version here), which becomes

$$\psi_+ = \left(\frac{m_{older}}{m_{older} + m_{younger}} \right)^{1/2} \psi_{younger} + \left(\frac{m_{younger}}{m_{older} + m_{younger}} \right)^{1/2} \psi_{older}$$

$$= \left(\frac{3}{4} \right)^{1/2} \psi_{1,1} + \left(\frac{1}{4} \right)^{1/2} \psi_{3,0} = \frac{1}{2} [\sqrt{3} \psi_{1,1} + \psi_{3,0}].$$

By hand calculator I sketched out this function ψ_+ on the interval $[0, 8)$ by finding it to be approximately: $\psi = 0.176$ on $[0, 2)$, 0.788 on $[2, 3)$, -0.436 on $[3, 4)$, and -0.0176 on $[4, 8)$. It produces an age turning angle of $30°$, because

$$\sin \phi(\psi_{1,1}, \psi_{3,0}) = \frac{m_{older} - m_{younger}}{m_{older} + m_{younger}} = \frac{3 - 1}{3 + 1} = \frac{1}{2}.$$

You should do the same for ψ_-.

2. See the bibliographies in our papers.
3. S was defined by $Sx = \sum_n \langle x, x_n \rangle x_n$ so first we must assure that S is an operator, i.e., that the series defining it converges. But the partial sums can be seen to be a Cauchy sequence:

$$\left\| \sum^N \langle x, x_n \rangle x_n - \sum^M \langle x, x_n \rangle x_n \right\|^2 = \left\| \sum_N^M \langle x, x_n \rangle x_n \right\|^2 \overset{\leq}{=} B \sum_N^M |\langle x, x_n \rangle|^2$$

which by the upper frame bound must go to zero. Then we may look at the quadratic form view of S, namely, $\langle Sx, x \rangle$, and see that

$$\langle Sx, x \rangle = \left\langle \sum_n \langle x, x_n \rangle x_n, x \right\rangle = \sum_n \langle x, x_n \rangle \langle x_n, x \rangle$$

$$= \sum_n |\langle x, x_n \rangle|^2.$$

Thus we may simply rewrite the original frame as

$$A\|x\|^2 \overset{\leq}{=} \langle Sx, x \rangle \overset{\leq}{=} B\|x\|^2.$$

From this by standard operator theory we know that

$$B^{-1}\|x\|^2 \overset{\leq}{=} \langle S^{-1}x, x \rangle \overset{\leq}{=} A^{-1}\|x\|^2.$$

But $\langle S^{-1}x, x \rangle = \sum_n |\langle x, x_n \rangle|^2$ by the above reasoning so $\{S^{-1}x_n\}$ is a frame.

4. It is better to think of S in its sesquilinear form bounds than in its operator bounds, although since S is SPD, they are equivalent. Recall our situation, stated here somewhat redundantly:

$$0 < A \overset{\leq}{=} \|S^{-1}\|^{-1} = m_S \overset{\leq}{=} S \overset{\leq}{=} M_S = \|S\| \overset{\leq}{=} B.$$

From this we have as operator forms, for any $\epsilon > 0$,

$$-I \overset{\leq}{=} \epsilon A - I \overset{\leq}{=} \epsilon m_S - I \overset{\leq}{=} \epsilon S - I \overset{\leq}{=} \epsilon M_S - I \overset{\leq}{=} \epsilon B - I.$$

To keep everything here positive so that we may automatically convert to operator norms, we impose lower bound $\epsilon > 1/A$. To keep everything bounded above by 1, we impose upper bound $\epsilon < 2/B$. For that interval $A^{-1} < \epsilon < 2B^{-1}$ we are assured the norm curve satisfies

$$0 < \|\epsilon S - I\| < 1.$$

However, we already know generally that $\|\epsilon S - I\|$ is less than one for all $0 < \epsilon < 2B^{-1}$, so the A^{-1} lower bound is not needed to keep $\|\epsilon S - I\|$ less than one. On the other hand we know that $\epsilon_m = 2/(m_S + M_S)$ exactly. So from the same form inequalities above we have the bound $A^{-1} < \epsilon_m < B^{-1}$ for the minimizing ϵ_m. Notice that if A and B were exact, then we would know that $\epsilon_m = 2/(A + B)$, so the bound $A^{-1} < \epsilon_m < B^{-1}$ then becomes $A > (A + B)/2 > B$, which is just the arithmetic mean inequality.

5. I leave it to the reader to play trigonometrically (in the conventional trigonometric sense) with some of these new Basis Trigonometry considerations and examples.

6. Dropping the normalization factor four for now, we have

$$\psi_+ = 3 \sin x - \sin 3x,$$

$$A\psi_+ = -\psi''_+ = 3\sin x - 9\sin 3x,$$

$$\|A\psi_+\|^2 = 9\|\sin x\|^2 + 81\|\sin 3x\|^2 = 45\pi,$$

$$\|A\psi_+\| = 3\pi^{1/2}\sqrt{5},$$

$$\|\psi_+\|^2 = 9\|\sin x\|^2 + \|\sin 3x\|^2 = 5\pi,$$

$$\|\psi_+\| = \pi^{1/2}\sqrt{5}.$$

Thus the antieigenvalue quotient $\mu(\psi_+)$ is

$$\mu(\psi_+) = \frac{\langle Ax, x\rangle}{\|Ax\|\|x\|} = \frac{9\|\sin x\|^2 + 9\|\sin 3x\|^2}{15\pi} = \frac{9\pi}{15\pi} = 0.6$$

which is $\cos(30°) = \cos\phi(A) = \mu_1(A)$. So ψ_+ attains the maximum turning angle $\psi(A)$.

There are other things one could check. But let us do the Euler equation

$$\frac{A^2 x}{\langle A^2 x, x\rangle} - \frac{2Ax}{\langle Ax, x\rangle} + x = 0.$$

For this we need $\|x\| = 1$, so we go back to normalized

$$\psi_+ = \frac{3\sin x - \sin 3x}{\sqrt{10}} \cdot \frac{1}{\|\sin nx\|} = \frac{1}{\sqrt{5\pi}}(3\sin x - \sin 3x).$$

From the above we already know

$$\langle A^2\psi_+, \psi_+\rangle = \|A\psi_+\|^2 = \frac{45\pi}{5\pi} = 9,$$

$$\langle A\psi_+, \psi_+\rangle = \frac{9\pi}{5\pi} = \frac{9}{5},$$

and we need

$$A^2\psi_+ = -(3\sin x - 9\sin 3x)'' \cdot \frac{1}{\sqrt{5\pi}}$$

$$= (3\sin x - 81\sin 3x) \cdot \frac{1}{\sqrt{5\pi}}.$$

We needed $\|\psi_+\| = 1$ in the $\langle A^2\psi_+, \psi_+\rangle$ and $\langle A\psi_+, \psi_+\rangle$ but a look at the Euler equation which is homogeneous in its numerators means we can now drop the $\sqrt{5\pi}$ and substitute the other quantities in, from which

$$\frac{A^2\psi_+}{\langle A^2\psi_+, \psi_+\rangle} - \frac{2A\psi_+}{\langle A\psi_+, \psi_+\rangle} + \psi_+$$

$$= \frac{3\sin x - 81\sin 3x}{9} - \frac{2(3\sin x - 9\sin 3x)}{9/5} + (3\sin x - \sin 3x)$$

$$= \frac{3\sin x - 81\sin 3x - 10(3\sin x - 9\sin 3x) + 27\sin x - 9\sin 3x}{9}$$

$$= \frac{\sin x[3 - 30 + 27] + \sin 3x[-81 + 90 - 9]}{9} = 0.$$

Chapter 6.

1. I found the unexpected (at least to me) connection to Von Neumann's consulting work at the Aberdeen Proving Grounds (as the Aberdeen laboratories were referred to when I worked at the Naval Research Laboratory in 1959–1963) just by following my curiosity. Therefore, you must follow your own curiosity to uncover historical facets of Ito's work on rocket-flight ballistics in Japan. These days, Google or other search engines will get you instant results. To possibly stoke some further curiosity, I mention that the Ito calculus will figure fundamentally in the financial instruments to which I will apply my operator trigonometry in Chapter 8 of this book.

2. Here is Von Neumann's matrix for $n = 3$:

$$A = \begin{bmatrix} 1 & -1 & 0 \\ -1 & 2 & -1 \\ 0 & -1 & 1 \end{bmatrix}.$$

To find its eigenvalues we set $\det(A - \lambda I) = 0$,

$$0 = \det(A - \lambda I) = (1 - \lambda)^2(2 - \lambda) - 2(1 - \lambda),$$

from which $\lambda = 0, 1, 3$. Checking this against Von Neumann's general formula $4\sin^2(k\pi/2n)$ for the eigenvalues, we see that the angles therein are 0, $30°$, and $60°$ and thus our λ's do check out.

Generally, the matrices

$$A = \begin{bmatrix} 2 & -1 & 0 & \cdots & 0 \\ -1 & 2 & -1 & \cdots & 0 \\ 0 & -1 & 2 & -1 & \vdots \\ \vdots & \ddots & \ddots & \ddots & -1 \\ 0 & \cdots & 0 & -1 & 2 \end{bmatrix}$$

are perhaps the most important matrices in numerical linear algebra, as they are the discretization by centered differences at all interior grid points for the one-dimensional Laplacian operator $-d^2/dx^2$. See my book [Gustafson (1999f)]. Von Neumann's statistically inspired matrix is the same except for its first and last rows, which cause the eigenvalue $\lambda = 0$ to appear in what otherwise would be a positive definite matrix like that above. Those first and last rows could come, in the PDE context, from Neumann boundary conditions implemented numerically by a first-order centered difference at left and right boundaries of the interval. It does not take much imagination to wonder how other simple ODE or PDE boundary value problems might induce interesting new statistics. That is the direction that seems to me the most interesting in this exercise.

I notice that these eigenvalue $4\sin^2(k\pi/2n)$ also surface in my treatment of numerical stability in my book [Gustafson (1999f), p. 312].

3. Bloomfield and Watson (1975) also used the Lagrange multiplier arguments I gave in Sec. 6.2 to arrive at the inefficiency equation. First they show that the matrices $X'\Gamma^2 X$ and $X'\Gamma X$ commute and thus may be diagonalized by the same orthogonal change of basis. Then they find the inefficiency equation in the form

$$\frac{Vx_i}{\langle Vx_i, x_i\rangle} + \frac{V^{-1}x_i}{\langle V^{-1}x_i, x_i\rangle} = 2x_i.$$

Here I have used the more common notation V rather than their Γ. The x_i denotes any column of the optimizing (worst efficiency) regression design matrix X. Multiplying this equation by V shows that anything in $\mathrm{sp}\{x_i, Vx_i\}$ is mapped into that span by V. Such two- (or one-) dimensional reducing subspace must have a two- (or one-) dimensional eigenbasis from V.

The specific coefficients for the inefficiency vectors and the antieigenvectors are then found from these 2×2 reduced subsystems as I showed in Sec. 6.2. It might be useful here to clarify two further points one could wonder about when reading their paper.

First is the jump to the quadratic equation

$$z^2 - 2\langle Vx, x\rangle z + \langle Vx, x\rangle\langle V^{-1}x, x\rangle^{-1} = 0.$$

One need not do that, i.e., one can just by brute force let $x = c_j x_j + c_k x_k$ and solve for c_j and c_k after substituting into the inefficiency equation. See how I did that in the Euler equation in Sec. 3.4. But it is faster to note that directly from the inefficiency equation, each of the eigenvalues λ_j and λ_k satisfies the scalar quadratic equation.

The second point is the jump from vectors x to matrices X one encounters in this kind of statistical estimation literature. The notation $|X'AX|$ means determinant, but what is meant by $X'AX$ itself? For me, it was useful to keep in mind its straightforward linear algebra content. Namely, consider the 2×2 case. Then write

$$
\begin{aligned}
X'AX &= \begin{bmatrix} [x_1^T] \\ [x_2^T] \end{bmatrix} [A][[x_1][x_2]] \\
&= \begin{bmatrix} [x_1^T] \\ [x_2^T] \end{bmatrix} [[Ax_1][Ax_2]] \\
&= \begin{bmatrix} \langle Ax_1, x_1 \rangle & \langle Ax_2, x_1 \rangle \\ \langle Ax_1, x_2 \rangle & \langle Ax_2, x_2 \rangle \end{bmatrix},
\end{aligned}
$$

which for $A = V$ clearly brings you to the correlation matrix context starting from individual vectors.

4. This is an open-ended research problem for someone other than me. First, one should do a literature search to come up-to-date on any following papers that might have appeared. Is the question obsolete now, from an engineering point of view? Whether that be the case or not, what new pure Rayleigh quotient/Euler equation mathematical theory can be developed, for possible use elsewhere in the future? Hints are the claim of angle bisection in [Cameron (1983)], and that the constraint trace (KK^T) be minimized in [Cameron and Kouvaritakis (1980)]. The latter is an SVD problem, asking for most efficient solution. Indeed the latter paper employs some Moore–Penrose theory. What does the maximization of $\cos^2 \theta_b \cos^2 \theta_c$ in [Cameron and Kouvaritakis (1980)] mean operator-theoretically?

5. See my paper [Gustafson (2000d)] where I do this.

6. Given the decomposition $X = M \oplus N$ of the linear space X, and P defined by $Px = m$ for each $x = m \oplus n$, one says the P projects on M along N. The geometry is that of projecting from the direction of N. In the example of the answer to Exercise 3 of Chapter 1, M is sp $\{x\}$ and N is sp $\{y^\perp\}$. As to trigonometry, of course P has its internal canonical angle

theory. I have not pushed an operator trigonometry onto P, but an interesting new one might come from its $P = QR$ factorization, using R in the spirit of my paper [Gustafson (2000d)]. There we noted that for any matrix A factored in both polar and QR forms, i.e., $A = U|A| = QR$, one has from A^*A that $|R| = |A|$ and hence my entended operator trigonometry applies to R as well. But my suggestion (admittedly vague) here is not that. I suppose I am wondering about the operator trigonometry of upper triangular matrices. One should even create a trigonometry coming just from Jordan canonical forms. How does one bring in the effect of the quasi-nilpotent shift part of $J(\lambda)$?

Chapter 7.

1. We wish to show that any a, b, c, all in the interval $[-1, 1]$, always satisfy the inequality

$$ab - bc + ac \overset{\leq}{=} 1.$$

Here is a proof. From $b^2 \overset{\leq}{=} 1$ and $c^2 \overset{\leq}{=} 1$ we have $b^2(1 - c^2) \overset{\leq}{=} 1 - c^2$ and hence $b^2 + c^2 \overset{\leq}{=} 1 + b^2c^2$. Adding $2bc$ to both sides and multiplying by $a^2 \overset{\leq}{=} 1$ we therefore have $a^2(b^2 + c^2 + 2bc) \overset{\leq}{=} b^2 + c^2 + 2bc \overset{\leq}{=} 1 + b^2c^2 + 2bc$, that is, $a^2(b + c)^2 \overset{\leq}{=} (1 + bc)^2$. Taking the positive square root yields $a(b + c) \overset{\leq}{=} |a||b + c| \overset{\leq}{=} 1 + bc$.

2. See [Gustafson (2000c)]. Also note the following minor typos in that paper: Theorem 2.2 on p. 38 assumes unit vectors, the second term in the inequality (17) on p. 40 is $1/2 \sin^2(\theta_{12})$, and on pp. 48 and 49 the factorization referred to is (50), and not (44), which is not a factorization at all.

A simple inequality made equality is just the Cauchy–Schwarz inequality

$$|\langle u, v \rangle| \overset{\leq}{=} \|u\|\|v\|$$

becoming

$$|\langle u, v \rangle| = \|u\|\|v\| \cos \theta_{u,v}.$$

Speaking more generally than just to my new quantum spin inequalities-made-equalities, I would enjoy very much the thought of an accepted legacy that whenever possible, inequalities should be extended to equalities throughout mathematics. Everyone likes to prove an inequality to be sharp by exhibiting an example in which equality is obtained. But why not go further and try to fill in all of the nonsharp region with meaningful new terms and thereby replace the full inequality with its inequality-equality?

3. A unitary matrix has its eigenvalues all on the unit circle in the complex plane. While that is great when you want to play with phase, it is defeating when you need differing amplitudes. You have only unit amplitudes for the eigenvalues of a unitary matrix. Stated more simply, we have seen that our operator angles are better thought of as twisting angles rather than just as turning angles.

4. The key is the condition number $\kappa = \lambda_2/\lambda_1$. One has

$$\frac{1 + \sin \phi(H)}{1 - \sin \phi(H)} = \frac{1 + \left(\frac{\lambda_2 - \lambda_1}{\lambda_2 + \lambda_1}\right)}{1 - \left(\frac{\lambda_2 - \lambda_1}{\lambda_2 + \lambda_1}\right)} = \frac{\lambda_2}{\lambda_1} = \kappa(H).$$

Note that when obtaining these new operator trigonometric identities, one also mixes in some of the customary trigonometry.

5. $M_\epsilon = U + \epsilon U^T$ has its two eigenvalues

$$\lambda_{1,2} = \frac{(1 + \epsilon) \pm i(1 - \epsilon)}{\sqrt{2}}$$

on the $(1 + \epsilon^2)^{1/2}$ circle. With $\epsilon > 0$, M_ϵ is still a strongly accretive normal matrix with numerical range $W(M_\epsilon)$ the convex hull of its two eigenvalues. Then, proceeding as in [Gustafson (2000d)], see Example 2 there, computation of the operator angle $\phi_{\text{re}}(M_\epsilon)$ follows from

$$\cos \phi_{\text{re}}(A) = \mu_1(A) = \min \left\{ \frac{\beta_k}{1 |\lambda_k|} \right\},$$

given the eigenvalues $\lambda_k = \beta_k + i\delta_k$. Thus we have

$$\phi_{\text{re}}(M_\epsilon) = \cos^{-1} \left(\frac{1 + 0.0023}{\sqrt{2}} \right) = 44.868°.$$

6. Let us develop the operator trigonometry of D. Immediately we find that the operator maximal turning angle and the first antieigenvalue are given by

$$\mu = \cos \phi(D) = \frac{2\sqrt{\lambda_1 \lambda_2}}{\lambda_1 + \lambda_2} = (1 - r)^{1/2}(1 + r)^{1/2}.$$

These thus depend only on the distance r from the center of the Bloch sphere. The eigenvectors are easily found and do not depend on r. But the antieigenvectors do: they are

$$x^{\pm} = \pm \left(\frac{1+r}{2}\right)^{1/2} \begin{bmatrix} 1 \\ e^{i\phi} \tan \frac{\theta}{2} \end{bmatrix} + \left(\frac{1-r}{2}\right)^{1/2} \begin{bmatrix} e^{-i\phi} \tan \frac{\theta}{2} \\ -1 \end{bmatrix}.$$

This is reminiscent of a behavior we found in twistors. There the eigenvectors did not depend on the time t, whereas the antieigenvectors did depend on t. So again the new trigonometry shows that it is the antieigenvectors which are needed to see the "twist" angles of the spin density states.

Chapter 8.

1. Part (a) follows trivially from the property for A an SPD matrix that

$$\sin \phi(A) = \frac{\lambda_1 - \lambda_n}{\lambda_1 + \lambda_n}.$$

I note we have used the statistician's ordering of λ_1 as the largest eigenvalue here. As for part (b) we have often used the property that the operator angle $\phi(A^{-1}) = \phi(A)$. Thus $\cos \phi(A^{-1}) = \cos \phi(A)$ and $\sin \phi(A^{-1}) = \sin \phi(A)$, and most of the operator trigonometry of A^{-1} is the same as that of A. However, the antieigenvectors of A^{-1} have the same weighting factors as the antieigenvectors of A, but the two eigenvectors switch roles. The proof consists of recalling first of all that A and A^{-1} have the same eigenvectors but inverted spectrum, so the eigenvector role-switch is clear. Then check that

$$\frac{\lambda_1(A^{-1})}{\lambda_1(A^{-1}) + \lambda_n(A^{-1})} = \frac{\lambda_1(A)}{\lambda_1(A) + \lambda_n(A)}.$$

In particular, note that A and A^{-1} have the same operator turning angles, the same eigenvectors, but not the same antieigenvectors.

2. We may equate the expression for $\lambda_1/(\lambda_1 + \lambda_2)$ found in Sec. 8.2 to that found in Lemma 8.4 above. Thus we have

$$\frac{1 + [1 - \cos^2 \phi(A_0) \sin^2 \psi(\rho)]^{1/2}}{2} = \frac{1 + \sin \phi(A)}{2}$$

from which, using $\sin^2 \phi(A) + \cos^2 \phi(A) = 1$, we have Lemma 8.1.

3. Originally, see [Gustafson (2010a)], I was in the enthralls of the entanglements of my operator trigonometry with the financial quantos instrument. But now I see that the trigonometry of Lemma 8.2 just by itself

is pretty trivial. To show the important (financially) implication of Lemma 8.2,

$$\rho\sigma_1\sigma_2\left[\frac{2}{\sigma_1^2 + \sigma_2^2}\right] = \cos\phi(A_0)\cos\psi(\rho) = \cos\phi(A)\cot\psi(\rho),$$

just for fun write the last desired equality as

$$\frac{\cos\phi(A)}{\cos\phi(A_0)} = \frac{\cos\psi(\rho)}{\cot\psi(\rho)}.$$

The left-hand side by the operator trigonometric cosine formula cancels out to $\sqrt{1 - \rho^2}$. That is the right-hand side which by elementary trigonometry is just $\sin\psi(\rho)$. To get the first equality, just note that the quantos drift term is

$$\frac{\rho}{\sqrt{1 - \rho^2}} \cdot \cos\phi(A)$$

which is the leftmost term. So it is all about correlations.

Going back to the financial context, do note that our scaling was by the operator trigonometric

$$\epsilon_m = \frac{2}{\sigma_1^2 + \sigma_2^2}$$

which gives you the convex minimum

$$\sin\phi(A_0) = \min_{\epsilon>0} \|\epsilon A_0 - I\|.$$

4. I still like, for practice, to Gauss-reduce small matrices to their inverses. But here the old adjugate formula is much quicker, from which

$$\Gamma^{-1} = \frac{\begin{bmatrix} \sigma_K^2 & -\sigma_2\sigma_K \\ -\sigma_2\sigma_K & \sigma_1^2 + \sigma_2^2 \end{bmatrix}}{\sigma_K^2\sigma_1^2} = \frac{1}{\sigma_1^2}\begin{bmatrix} 1 & -\frac{\sigma_2}{\sigma_K} \\ -\frac{\sigma_1}{\sigma_K} & \frac{\sigma_1^2+\sigma_2^2}{\sigma_K^2} \end{bmatrix}.$$

Then

$$\Gamma^{-1}\begin{bmatrix} \sigma_1\lambda + \sigma_2\lambda_r \\ \sigma_K\lambda_r \end{bmatrix} = \frac{1}{\sigma_1^2}\begin{bmatrix} \sigma_1\lambda \\ -\frac{\sigma_2\sigma_1\lambda}{\sigma_K} + \frac{\sigma_1^2\lambda_r}{\sigma_K} \end{bmatrix} = \begin{bmatrix} \frac{\lambda}{\sigma} \\ -\frac{\lambda\sigma_2+\lambda_r\sigma_1}{\sigma_1\sigma_K} \end{bmatrix}.$$

5. See the Epilogue in my book [Gustafson (1999f)] and also [Merton (1990)] and especially [Wilmott (1998)].

6. Hint: see *American Mathematical Monthly* (1968), where I published a paper on compact operators. Thorp also played with those.

Chapter 9.

1. $B = (SS^*)^{-1}$ has adjoint

$$B^* = ((SS^*)^{-1})^* = ((SS^*)^*)^{-1} = (S^{**}S^*)^{-1} = (SS^*)^{-1}$$

where we used $S^{**} = S$ for any closed operator in a Hilbert space. See [Gustafson (2004b)] for the details about

$$A' = B^{-1}A^*B$$

being A's dual in the new inner product space. It can be tricky and that paper reveals a serious error in some preceding papers in the numerical linear algebra literature.

By the way, since as shown in Sec. 9.1 we have

$$\langle Ax, x \rangle_B = \langle Dy, y \rangle, \quad y = S^{-1}x,$$

you may check that similarly we have

$$\frac{|\langle Ax, x \rangle_B|}{\|Ax\|_B \|x\|_B} = \frac{|\langle Dy, y \rangle|}{\|Dy\| \|y\|}.$$

Thus the total antieigenvalue theory for A in the B-inner product is essentially (up to antieigenvectors) that of D in the original space.

2. U is not normal. However its eigenvalues and eigenvectors are easily calculated and one arrives at its diagonal representation

$$A = \begin{bmatrix} -\frac{1}{2} + i\frac{\sqrt{3}}{2} & 0 & 0 & 0 \\ 0 & -\frac{1}{2} - i\frac{\sqrt{3}}{2} & 0 & 0 \\ 0 & 0 & 1 & 0 \\ 0 & 0 & 0 & 0 \end{bmatrix}.$$

What I found equally interesting in [Gustafson (2011a)] was that A's normal polynomial, that is, the polynomial $n(\lambda)$ such that A^* is represented as $A^* = n(A)$, is also informationally deficient: $A^* = A^2$. Usually, in fact almost always, one finds $A^* = n(A)$ where $n(\lambda)$ has degree $n - 1$ for a given $n \times n$ matrix A. Surprisingly, the loss of terms in $n(\lambda)$ can be related to gravitational lensing. This phenomenon will be treated in a paper, Gustafson, *Normal Degree Revisited*, to appear.

3. This is fun. We can also do some variations. For $x = x_1 + x_n$ and $y = x_1 - x_n$ we obtain straightforwardly

$$\frac{\langle Ax, Ay \rangle}{\|Ax\| \|Ay\|} = \frac{\lambda_1^2 - \lambda_n^2}{\lambda_1^2 + \lambda_n^2}.$$

The right-hand side is $\sin \phi(A^2)$. However, we may for the moment ignore that and just notice that from conventional trigonometric identities, we have from the definition of the K–W angle from the conventional condition number κ according to

$$\cot \left(\frac{\theta(A)}{2} \right) = \kappa$$

that

$$\cos \theta(A) = \frac{\kappa^2 - 1}{\kappa^2 + 1} = \frac{\lambda_n^2 - \lambda_1^2}{\lambda_1^2 + \lambda_1^2}.$$

Thus $x = x_1 + x_n$ and $y = x_1 + x_n$ are optimal. They can of course be made of norm one due to the homogeneity of the K–W inequality entities.

Let us show here a variation employing instead the general operator trigonometric identity (see Lemma 4.1 in Sec. 4.3)

$$\sin \phi(A^{1/2}) = \frac{\sin \phi(A)}{1 + \cos \phi(A)}.$$

We have, continuing on from the above,

$$\frac{\langle Ax, Ay \rangle}{\|Ax\| \|Ay\|} = \frac{\lambda_1^2 - \lambda_n^2}{\lambda_1^2 + \lambda_n^2} = \left(\frac{\lambda_1 - \lambda_n}{\lambda_1 + \lambda_n} \right) \frac{(\lambda_1 + \lambda_n)^2}{(\lambda_1^2 + \lambda_n^2)}$$

$$= \sin \phi(A) \left[1 + \frac{2\lambda_1 \lambda_n}{\lambda_1^2 + \lambda_n^2} \right]$$

$$= \sin \phi(A)[1 + \cos \phi(A^2)]$$

$$= \frac{\sin \phi(A^2)}{[1 + \cos \phi(A^2)]} \cdot [1 + \cos \phi(A^2)]$$

and we are back again to the result that the K–W inequality is optimized to either $\cos \theta(A)$ from the original Kantorovich–Wielandt standard condition number view, or at $\sin \phi(A^2)$ from my view. The merit in this instance of the latter viewpoint is my point in Sec. 9.3 that my $\sin\phi(A)$ convex minimum, applied in this case to the $\sin\phi(A^2)$ minimum, contains the Kantorovich–Wielandt sharp vectors $x = x_1 + x_n$ and $y = x_1 - x_n$.

4. We may consider x_+, the case of x_- following in the same way. For convenience write

$$x_+ = \frac{\lambda_1^{1/4} x_1 + \lambda_n^{1/4} x_n}{d}, \qquad d = (\lambda_1^{1/2} + \lambda_n^{1/2}).$$

Then

$$\langle Ax, x\rangle - \langle A^{-1}x, x\rangle^{-1} = \frac{\lambda_1^{3/2} + \lambda_n^{3/2}}{d^2} - \frac{d^2}{\lambda_1^{-1/2} + \lambda_n^{-1/2}}.$$

Using the factorization $(a+b)^3 = (a+b)(a^2 - ab + b^2)$, the first term is $\lambda_1 - \lambda_1^{1/2}\lambda_n^{1/2} + \lambda_n$. Thus we need another $\lambda_1^{1/2}\lambda_n^{1/2}$ from the second term, which follows by elementary algebra.

The fact that these optimal $x_\pm$ are the antieigenvectors for the operator $A^{-1/2}$ means that maximizing the Shisha–Mond expression is equivalent to minimizing the expression

$$\frac{\langle A^{-1/2}x, x\rangle}{\|A^{-1/2}x\|\|x\|}.$$

It would be nice to work out the full details relating the two theories.

5. Writing the detail in vector component form, we have

$$x_+ = (\lambda_n^{1/2}, 0, \ldots, 0, \lambda_1^{1/2})(\lambda_1 + \lambda_n)^{-1/2},$$

$$Ax_+ = (\lambda_n^{1/2}\lambda_1, 0, \ldots, 0, \lambda_1^{1/2}\lambda_n)(\lambda_1 + \lambda_n)^{-1/2}$$

and we recall that $\|Ax_\pm\| = \lambda_n^{1/2}\lambda_1^{1/2}$. For $y = Ax_+$ the slack variable numerator of the expression for $\sin\phi(A)$ in Theorem 9.1 is, writing everything out in that ij sum to be absolutely clear,

$$(x_1y_2 - y_1x_2)^2 + (x_1y_3 - y_1x_3^2) + \cdots + (x_1y_n - y_1x_n)^2$$

$$+ (x_2y_3 - y_2x_3)^2 + \cdots + (x_2y_n - y_2x_n)^2 \cdots$$

$$+ (x_{n-1}y_n - y_{n-1}x_n)^2$$

$$= (O)^2 + (O)^2 + \cdots + (\lambda_n^{1/2}\lambda_1^{1/2}\lambda_n - \lambda_n^{1/2}\lambda_1\lambda_1^{1/2})^2 \cdot (\lambda_1 + \lambda_n)^{-2}$$

$$+ (O)^2 + \cdots + \cdots + (O)^2 \cdots + (O)^2.$$

6. (a) The first antieigenvalue is

$$\mu_1 = \cos\phi(A) = \frac{2\lambda^2}{\lambda^4 + 1} = \frac{2(1 + \lambda)}{(2 + 3\lambda) + 1} = \frac{2}{3}.$$

Here I found it fun to play with the golden mean λ, from which one stumbles into the reductions

$$\lambda^2 = 1 + \lambda, \quad \lambda^4 = 2 + 3\lambda, \quad \lambda^6 = 5 + 8\lambda, \quad \lambda^3 = 1 + 2\lambda.$$

Hence A's operator angle is $\phi(A) \cong 48.1896851°$. Also we have independently

$$v_1 = \sin\phi(A) = \frac{\lambda^4 - 1}{\lambda^4 + 1} = \frac{3\lambda + 1}{3\lambda + 3} = \frac{1}{3} + \frac{2}{3}\left(\frac{1}{\lambda}\right).$$

Thus we may check that $\sin^2\phi(A) + \cos^2\phi(A) = 1$ by checking

$$\left(\frac{1}{3} + \frac{2}{3}\left(\frac{1}{\lambda}\right)\right)^2 + \frac{4}{9} = \frac{1}{9}\left[5 + 4\left(\frac{1}{\lambda^2} + \frac{1}{\lambda}\right)\right] = \frac{9}{9}.$$

Next we may form the antieigenvectors

$$x_{\pm} = \left(\frac{\lambda^2}{\sqrt{\lambda^4 + 1}}\right) \cdot \frac{1}{\sqrt{\lambda^2 + 1}}\begin{bmatrix} -1 \\ \lambda \end{bmatrix} \pm \left(\frac{1}{\sqrt{\lambda^4 + 1}}\right) \cdot \frac{1}{\sqrt{\lambda^2 + 1}}\begin{bmatrix} \lambda \\ 1 \end{bmatrix}.$$

Let

$$d = \sqrt{\lambda^4 + 1} \cdot \sqrt{\lambda^2 + 1} = (\lambda^6 + \lambda^4 + \lambda^2 + 1)^{1/2}$$
$$= (12\lambda + 9)^{1/2} \cong 5.330704256.$$

Then

$$x_+ = \frac{1}{d}\begin{bmatrix} -\lambda^2 + \lambda \\ \lambda^3 + 1 \end{bmatrix} = \frac{1}{d}\begin{bmatrix} -1 \\ 2\lambda + 2 \end{bmatrix} \cong \begin{bmatrix} -0.1876 \\ 0.9824 \end{bmatrix},$$

$$x_- = \frac{1}{d}\begin{bmatrix} -\lambda^2 - \lambda \\ \lambda^3 - 1 \end{bmatrix} = \frac{1}{d}\begin{bmatrix} -1 - 2\lambda \\ 2\lambda \end{bmatrix} \cong \begin{bmatrix} -0.7948 \\ 0.3686 \end{bmatrix}.$$

Notice

$$\|x_+\|^2 = \frac{1 + 4\lambda^2 + 4 + 8\lambda}{12\lambda + 9} = \frac{12\lambda + 9}{12\lambda + 9} = 1.$$

Similarly, as a check, we have

$$\|x_-\|^2 = \frac{1 + 4\lambda + 8\lambda^2}{12\lambda + 9} = 1.$$

(b) In the A-inner product we have

$$\langle x_+, x_- \rangle_A \equiv \langle Ax_+, x_- \rangle = \frac{\lambda_n \lambda_1}{\lambda_1 + \lambda_n}[\langle x_1, x_1 \rangle - \langle x_n, x_n \rangle] = 0.$$

So the full antieigenvector set $\{x_\pm\}$ becomes an orthonormal basis in the A-inner product.

(c) We may compute the A inner product

$$\langle Ax_+, x_- \rangle = \frac{\left\langle \begin{bmatrix} 2 & 1 \\ 1 & 1 \end{bmatrix} \begin{bmatrix} -1 \\ 2\lambda+1 \end{bmatrix}, \begin{bmatrix} -1-2\lambda \\ 2\lambda \end{bmatrix} \right\rangle}{(12\lambda+9)}$$

$$= \frac{[-1-2\lambda, 2\lambda]}{(12\lambda+9)} \begin{bmatrix} 2\lambda \\ 2\lambda+1 \end{bmatrix}$$

$$= \frac{-2\lambda - 4\lambda^2 + 4\lambda^2 + 2\lambda}{12\lambda+9} = 0.$$

In the original inner product we have

$$\langle x_+, x_- \rangle = \frac{(1+2\lambda) + (2\lambda+2)(2\lambda)}{(12\lambda+9)}$$

$$= \frac{1+6\lambda+4\lambda^2}{12\lambda+9} = \frac{10\lambda+5}{12\lambda+9}$$

$$= \frac{10+5\sqrt{5}}{15+6\sqrt{5}} = \frac{21.18033989}{28.41640787}$$

$$= 0.74535599 = \sin\phi(A).$$

Bibliography

Bib. 1 lists the author's contributions (sometimes with coauthors) which have operator-trigonometric content. Bib. 2 lists what he deems important contributions by others which have direct, or indirect, operator-trigonometric content. Bib. 3 lists the remaining bibliography which were cited in this book.

Bib. 1 Contributions by K. Gustafson (and Coauthors)

K. Gustafson. (1966). A Perturbation Lemma, *Bull. Amer. Math. Soc.* **72**, pp. 334–338.

K. Gustafson. (1967). Positive Operator Products, *Notices Amer. Math. Soc.* **14**, Abstract 67T-531, p. 717. See also Abstracts 67T-340, 67T-564, 67T-675.

K. Gustafson. (1968a). A Note on Left Multiplication of Semi-group Generators, *Pacific J. Math.* **24**, pp. 463–465.

K. Gustafson. (1968b). The Angle of an Operator and Positive Operator Products, *Bull. Amer. Math. Soc.* **74**, pp. 488–492.

K. Gustafson. (1968c). Positive (noncommuting) Operator Products and Semi-groups, *Math. Zeitschrift* **105**, pp. 160–172.

K. Gustafson. (1968d). A Min–Max Theorem, *Amer. Math. Soc. Notices* **15**, p. 799.

K. Gustafson. (1969a). Doubling Perturbation Sizes and Preservation of Operator Indices in Normed Linear Spaces, *Proc. Camb. Phil. Soc.* **98**, pp. 281–294.

K. Gustafson. (1969b). On the Cosine of Unbounded Operators, *Acta Sci. Math.* **30**, pp. 33–34 (with B. Zwahlen).

K. Gustafson. (1969c). Some Perturbation Theorems for Nonnegative Contraction Semi-Groups, *J. Math. Soc. Japan* **21**, pp. 200–204 (with Ken-iti Sato).

K. Gustafson. (1970a). A Simple Proof of the Toeplitz–Hausdorff Theorem for Linear Operators, *Proc. Amer. Math. Soc.* **25**, pp. 203–204.

K. Gustafson. (1972a). Anti-eigenvalue Inequalities in Operator Theory, *Inequalities III*, Shisha, O. (ed.), (Academic Press), pp. 115–119.

K. Gustafson. (1972b). Multiplicative Perturbation of Semigroup Generators, *Pac. J. Math.* **41**, pp. 731–742 (with G. Lumer).

K. Gustafson. (1972c). Multiplicative Perturbation of Nonlinear m-Accretive Operators, *J. Funct. Anal.* **10**, pp. 149–158 (with B. Calvert).

K. Gustafson. (1977a). Numerical Range and Accretivity of Operator Products, *J. Math. Anal. Applic.* **60**, pp. 693–702 (with D. Rao).

K. Gustafson. (1983a). The RKNG (Rellich, Kato, Nagy, Gustafson) Perturbation Theorem for Linear Operators in Hilbert and Banach Space, *Acta Sci. Math.* **45**, pp. 201–211.

K. Gustafson. (1989a). Antieigenvalue Bounds, *J. Math. Anal. Applic.* **143**, pp. 327–340 (with M. Seddighin).

K. Gustafson. (1991a). Antieigenvalues in Analysis, in *Fourth International Workshop in Analysis and its Applications*, Stanojevic, C. and Hadzic, O. (eds.), Dubrovnik, Yugoslavia, June 1–10, 1990 (Novi Sad, Yugoslavia), pp. 57–69.

K. Gustafson. (1993a). A Note on Total Antieigenvectors, *J. Math. Anal. Applic.* **178**, pp. 603–611 (with M. Seddighin).

K. Gustafson. (1994a). Operator Trigonometry, *Linear and Multilinear Algebra* **37**, pp. 139–159.

K. Gustafson. (1994b). Antieigenvalues, *Lin. Alg. and Applic.* **208/209**, pp. 437–454.

K. Gustafson. (1994c). Computational Trigonometry, *Proc. Colorado Conf. on Iterative Methods*, Vol. 1, p. 1.

K. Gustafson. (1995a). Matrix Trigonometry, *Lin. Alg. and Applic.* **217**, pp. 117–140.

K. Gustafson. (1996a). *Lectures on Computational Fluid Dynamics, Mathematical Physics, and Linear Algebra* (Kaigai Publishers, Tokyo, Japan).

K. Gustafson. (1996b). Trigonometric Interpretation of Iterative Methods, in *Proc. Conf. Algebraic Multilevel Iteration Methods with Applications*, Axelsson, O. and Polman, B. (eds.), (Nijmegen, Netherlands), June 13–15, pp. 23–29.

K. Gustafson. (1996c). Commentary on Topics in the Analytic Theory of Matrices, Section 23, Singular Angles of a Square Matrix, in *Collected Works of Helmut Wielandt 2*, Huppert, B. and Schneider, H. (eds.), (De Gruyters, Berlin), pp. 356–367.

K. Gustafson. (1997a). *Numerical Range: The Field of Values of Linear Operators and Matrices* (Springer, Berlin), (with D. Rao).

K. Gustafson. (1997b). Operator Trigonometry of Iterative Methods, *Num. Lin. Alg. with Applic.* **34**, pp. 333–347.

K. Gustafson. (1997c). Antieigenvalues, *Encyclopaedia of Mathematics, Supplement 1*, (Kluwer Acad. Publ., Dordrecht), p. 57.

K. Gustafson. (1997d). *Lectures on Computational Fluid Dynamics, Mathematical Physics, and Linear Algebra*, (World Scientific, Singapore).

K. Gustafson. (1998a). Domain Decomposition, Operator Trigonometry, Robin Condition, *Contemp. Math.* **218**, pp. 455–460.

K. Gustafson. (1998b). Operator Trigonometry of Wavelet Frames, in *Iterative Methods in Scientific Computation*, Wang, J., Allen, M., Chen, B., Mathew, T. (eds.), *IMACS Series in Computational and Applied Mathematics*, Vol. 4 (New Brunswick, NJ), pp. 161–166.

K. Gustafson. (1998c). Semigroups and Antieigenvalues, in *Irreversibility and Causality — Semigroups and Rigged Hilbert Spaces*, Bohm, A., Doebner, H. and Kielanowski, P. (eds.), *Lecture Notes in Physics*, Vol. 504, (Springer, Berlin), pp. 379–384.

K. Gustafson. (1998d). Operator Trigonometry of Linear Systems, in *Proc. 8th IFAC Symposium on Large Scale Systems: Theory and Applications*, Koussoulas, N. and Groumpos, P. (eds.), Patras, Greece, July 15–17, (Pergamon Press, 1999), pp. 950–955.

K. Gustafson. (1998e). Symmetrized Product Definiteness? Comments on Solutions 19-5.1–19-5.5, *IMAGE: Bull. Int. Linear Algebra Soc.* **21**, p. 22.

K. Gustafson. (1998f). Antieigenvalues: An Extended Spectral Theory, in *Generalized Functions, Operator Theory and Dynamical Systems*, Antoniou, I. and Lumer, G. (eds.), *Pitman Research Notes in Mathematics*, Vol. 399, London, pp. 144–149.

K. Gustafson. (1998g). Operator Trigonometry of the Model Problem, *Num. Lin. Alg. with Applic.* **5**, pp. 377–399.

K. Gustafson. (1999a). The Geometry of Quantum Probabilities, in *On Quanta, Mind, and Matter: Hans Primas in Context*, Atmanspacher, H., Amann, A., and Mueller-Herold, U. (eds.), (Kluwer, Dordrecht), pp. 151–164.

K. Gustafson. (1999b). The Geometrical Meaning of the Kantorovich–Wielandt Inequalities, *Lin. Alg. and Applic.* **296**, pp. 143–151.

K. Gustafson. (1999c). Symmetrized Product Definiteness: A Further Comment, *IMAGE: Bull. Int. Linear Algebra Soc.* **22**, p. 26.

K. Gustafson. (1999d). A Computational Trigonometry and Related Contributions by Russians Kantorovich, Krein, Kaporin, *Computational Technologies* **4**(3), pp. 73–83, (Novosibirsk, Russia).

K. Gustafson. (1999e). *On Geometry of Statistical Efficiency*, preprint.

K. Gustafson. (2000a). The Trigonometry of Quantum Probabilities, in *Trends in Contemporary Infinite Dimensional Analysis and Quantum Probability*, Accardi, L., Kuo, H., Obata, N., Saito, K., Si, S., and Streit, L. (eds.), (Italian Institute of Culture, Kyoto), pp. 159–173.

K. Gustafson. (2000b). Semigroup Theory and Operator Trigonometry, in *Semigroups of Operators: Theory and Applications*, Balakrishnan, A.V. (ed.), (Birkhäuser, Basel), pp. 131–140.

K. Gustafson. (2000c). Quantum Trigonometry, *Inf. Dim. Anal. Quant. Probab. Relat. Top.* **3**, pp. 33–52.

K. Gustafson. (2000d). An Extended Operator Trigonometry, *Lin. Alg. and Applic.* **319**, pp. 117–135.

K. Gustafson. (2001a). An Unconventional Linear Algebra: Operator Trigonometry, in *Unconventional Models of Computation*, UMC'2K, Antoniou, I., Calude, C., and Dinneen, M. (eds.), (Springer, London), pp. 48–67.

K. Gustafson. (2001b). Probability, Geometry, and Irreversibility in Quantum Mechanics, *Chaos, Solitons and Fractals* **12**, pp. 2849–2858.

K. Gustafson. (2002a). Operator Trigonometry of Statistics and Econometrics, *Lin. Alg. and Applic.* **354**, pp. 151–158.

K. Gustafson. (2002b). CP-Violation as Antieigenvector-Breaking, *Adv. Chem. Phys.* **122**, pp. 239–258.

K. Gustafson. (2003a). Operator Trigonometry of Preconditioning, Domain Decomposition, Sparse Approximate Inverses, Successive Overrelaxation, Minimum Residual Schemes, *Num. Lin. Alg. with Applic.* **10**, pp. 291–315.

K. Gustafson. (2003b). Bell's Inequalities, in *Contributions to the XXII Solvay Conference on Physics*, Borisov, A. (ed.), (Moscow–Izhevsk, ISC, Moscow State University), pp. 501–517.

K. Gustafson. (2003c). Bell's Inequality and the Accardi–Gustafson Inequality, in *Foundations of Probability and Physics-2*, Khrennikov, A. (ed.), (Växjo University Press, Sweden), pp. 207–223.

K. Gustafson. (2003d). Bell's Inequalities, in *The Physics of Communication*, Proceedings of the XXII Solvay Conference on Physics, Antoniou, I., Sadovnichy, V., and Walther, H. (eds.), (World Scientific), pp. 534–554.

K. Gustafson. (2003e). in *Preconditioning, Inner Products, Normal Degree*, 2003 International Conference on Preconditioning Techniques for Large Sparse Matrix Problems in Scientific and Industrial Applications, Ng, E., Saad, Y., and Tang, W. P. (eds.), (NAPA, CA, October), 3 pp.

K. Gustafson. (2004a). An Inner Product Lemma, *Num. Lin. Alg. and Applic.* **11**, pp. 649–659.

K. Gustafson. (2004b). Normal Degree, *Num. Lin. Alg. and Applic.* **11**, pp. 661–674.

K. Gustafson. (2004c). Interaction Antieigenvalues, *J. Math. Anal. Applic.* **299**, pp. 174–185.

K. Gustafson. (2005a). The Geometry of Statistical Efficiency, *Res. Lett. Inf. Math. Sci.* **8**, pp. 105–121.

K. Gustafson. (2005b). On the Eigenvalues which Express Antieigenvalues, *Int. J. Math. Math. Sci.* **10**, pp. 1543–1554 (with M. Seddighin).

K. Gustafson. (2005c). Bell and Zeno, *Int. J. Theor. Phys.* **44**, pp. 1931–1940.

K. Gustafson. (2006a). Noncommutative Trigonometry, in *Operator Theory: Advances and Applications* **167**, pp. 127–155.

K. Gustafson. (2006b). The Trigonometry of Matrix Statistics, *Int. Statist. Rev.* **74**, pp. 187–202.

K. Gustafson. (2007a). Noncommutative Trigonometry and Quantum Mechanics, in *Advances in Deterministic and Stochastic Analysis*, Chuong N., Ciarlet P., Lax P., Mumford D. and Phong D. (eds.), (World Scientific), pp. 341–360.

K. Gustafson. (2007b). The Geometry of Statistical Efficiency and Matrix Statistics, *J. Appl. Math. Decis. Sci.*, doi:10.1155/2007/94515.

K. Gustafson. (2008a). The Operator Trigonometry in Statistics, in *Functional and Operatorial Statistics*, Dabo-Niang S. and Ferraty F. (eds.), pp. 189–193.

K. Gustafson. (2009a). The Trigonometry of Twistors and Elementary Particles, in *Foundations of Probability and Physics–5*, Accardi L., Adenier G., Fuchs C. A., Jaeger G., Khrennikov A. Yu., Larsson, J. A. and Stenholm S. (eds.), Amer. Inst. of Physics Conference Proceedings, Vol. 1101(AIP), pp. 65–73.

K. Gustafson. (2009b). Operator Trigonometry of Hotelling Correlation, Frobenius Condition, Penrose Twistor, *Lin. Alg. and Applic.* **430**, pp. 2762–2770.

K. Gustafson. (2010a). Operator Trigonometry of Multivariate Finance, *J. Multivar. Anal.* **101**, pp. 374–384.

K. Gustafson. (2010b). Slant Antieigenvalues and Slant Antieigenvectors, *Lin. Alg. and Applic.* **432**, pp. 1348–1362 (with M. Seddighin).

K. Gustafson. (2010c). A Trigonometry of Quantum States, in *Quantum Theory: Reconsideration of Foundations–5*, Khrennikov A. (ed.), Amer. Inst. of Physics Conference Proceedings, Vol. 1232(AIP), pp. 72–85.

K. Gustafson. (2010d). On my Min-Max Theorem (1968) and its Consequences, *Acta Comment. Univ. Tartu. Math.* **14**, pp. 45–51.

K. Gustafson. (2010e). *Operator Geometry of Statistics*, The Oxford Handbook of Functional Data Analysis, Ferraty F. and Romain Y. (eds.), (Oxford University Press), pp. 355–382.

K. Gustafson. (2011a). Trigonometry of Quantum States, *Found. Phys.* **41**, 450–465.

K. Gustafson. (2011b). Forty Years of Antieigenvalue Theory and Applications, Lecture given at IWMS 2010, Shanghai, China, preprint.

K. Gustafson. (2011c). Geometric Invariants and Operator Trigonometry, in preparation.

K. Gustafson. (2012a). The Financial Sharpe Ratio Seen as a New Risk Ratio via Antieigenvalue Theory, in preparation.

Bib. 2   Important Contributions by Others

L. Accardi and A. Fedullo. (1982). On the Statistical Meaning of Complex Numbers in Quantum Mechanics, *Lett. Nuovo. Cim.* **34**, pp. 161–172.

E. Asplund and V. Ptak. (1971). A Minmax Inequality for Operators and a Related Numerical Range, *Acta Math.* **126**, pp. 53–62.

C. Davis. (1980). Extending the Kantorovich Inequalities to Normal Matrices, *Lin. Alg. and Appl.* **31**, pp. 173–177.

S. Drury, S. Liu, C. Y. Lu, S. Puntanen and G. P. H. Styan. (2002). Some Comments on Several Matrix Inequalities with Applications to Canonical Correlations: Historical Background and Recent Developments, *Sankhya* **64**, Series A, Pt. 2, pp. 453–507.

P. Hess. (1971). A Remark on the Cosine of Linear Operators, *Acta Sci. Math.* **32**, pp. 267–269.

S. Hossein, K. Paul, L. Debnath and K. Das. (2008). Symmetric Antieigenvalue and Symmetric Antieigenvector, *J. Math. Anal. Applic.* **345**, pp. 771–776.

L. Kantorovich. (1948). Functional Analysis and Applied Mathematics, *Uspekhi Mat. Nauk* **3**(6), pp. 89–185.

R. Khatree. (2001). On Calculation of Antieigenvalues and Antieigenvectors, *J. Interdiscip. Math.* **4**, pp. 195–199.

R. Khatree. (2002). On Generalized Antieigenvalue and Antieigenmatrix of order *r*, *Amer. J. of Math. Management Sci.* **22**, pp. 89–98.

R. Khatree. (2003). Antieigenvalues and Antieigenvectors in Statistics, *J. Statist. Plan. Inference* **114**, pp. 131–144.

M. G. Krein. (1969). Angular Localization of the Spectrum of a Multiplicative Integral in a Hilbert Space, *Funct. Anal. Appl.* **3**, pp. 89–90.

B. A. Mirman. (1983). Antieigenvalues: Method of Estimation and Calculation, *Lin. Alg. and Appl.* **49**, pp. 247–255.

C. R. Rao. (2005). Antieigenvalues and Antisingularvalues of a Matrix and Application to Problems in Statistics, *Res. Lett. Inf. Math. Sci.* **8**, pp. 53–76.

C. R. Rao. (2007). Antieigenvalues and Antisingularvalues of a Matrix and Application to Problems in Statistics, *Math. Inequal. Appl.* **10**, pp. 471–489.

D. Rao. (1972). Numerical Range and Positivity of Operator Products, Ph.D. Dissertation, University of Colorado, Boulder, Colorado.

M. Seddighin. (2002). Antieigenvalues and Total Antieigenvalues of Normal Operators, *J. Math. Anal. Appl.* **274**, pp. 239–254.

M. Seddighin. (2005). On the Joint Antieigenvalue of Operators on Normal Subalgebras, *J. Math. Anal. Applic.* **312**, pp. 61–71.

M. Seddighin. (2009). Antieigenvalue Techniques in Statistics, *Lin. Alg. and Applic.* **430**, pp. 2566–2580.

M. Seddighin. (2011). Slant Joint Antieigenvalues and Antieigenvectors of Operators in Normal Subalgebras, *Lin. Alg. and Applic.* **434**, pp. 1395–1408.

H. Wielandt. (1967). Topics in the Analytic Theory of Matrices, *University of Wisconsin Lecture Notes*, Madison, Wisconsin.

Bib. 3   Other General and Specific Cited References

A. Afriat and F. Seleri. (1999). *The Einstein, Podolski, and Rosen Paradox*, (Plenum Press, New York).

J. Andreason, B. Jensen and R. Poulsen. (1998). Eight Valuation Methods in Financial Mathematics: The Black–Scholes Formula as an Example, *Math. Scientist* **23**, pp. 18–40.

A. Aspect, J. Dalibard and G. Roger. (1982). Experimental Test of Bell's Inequalities Using Time-Varying Analyzers, *Phys. Rev. Lett.* **49**, pp. 1804–1807.

J. Authers. (2010). *The Fearful Rise of Markets*, (Financial Times Press).

O. Axelsson. (1994). *Iterative Solution Methods*, (Cambridge Univ. Press).

I. Bajeux-Besnainou and R. Portait. (2002). Dynamic, Deterministic and Static Optimal Portfolio Strategies in a Mean-Variance Framework Under Stochastic Interest Rates, in Wilmott P. and Rasmussen H. (eds.), *New Directions in Mathematical Finance*, (John Wiley & Sons Ltd., London), pp. 101–115.

M. Baxter and A. Rennie. (1996). *Financial Calculus*, (Cambridge University Press, Cambridge, UK).

J. Bell. (1964). On the Einstein–Podolsky–Rosen Paradox, *Physics* **1**, pp. 195–200.

I. Bengtsson and K. Zyczkowski. (2008). *Geometry of Quantum States*, (Cambridge Univ. Press, Cambridge).

J. Bernasconi and K. Gustafson. (1992). Human and Machine "Quick Modelling". *Adv. Neural Info. Process. Syst.* **4**, pp. 1151–1158.

J. Bernasconi and K. Gustafson. (1994). Inductive Inference and Neural Nets. *Network: Comput. Neural Syst.* **5**, pp. 203–228.

J. Bernasconi and K. Gustafson. (1998). Contextual Quick-Learning and Generalization by Humans and Machines. *Network: Comput. Neural Syst.* **9**, pp. 85–106.

P. Bloomfield and G. S. Watson. (1975). The Inefficiency of Least Squares, *Biometrika* **62**, pp. 121–128.

U. Brehm. (1990). The Shape Invariant of Triangles and Trigonometry in Two-Point Homogeneous Spaces, *Geom. Dedicata* **33**, pp. 59–76.

R. Cameron and B. Kouvaritakis. (1980). Minimizing the Norm of Output Feedback Controllers Used in Pole Placement: A Dyadic Approach, *Int. J. Control* **32**, pp. 759–770.

R. Cameron. (1983). Minimizing the Product of Two Rayleigh Quotients, *Linear and Multilinear Algebra* **13**, pp. 177–778.

R. Carmona and V. Durrleman. (2003). Pricing and Hedging Spread Options, *SIAM Review* **45**, pp. 627–685.

P. Carr, H. Geman and D. Madan. (2001). Pricing and Hedging in Incomplete Markets, *J. Financial Economics* **62**, pp. 131–167.

F. Chaitin-Chatelin and S. Gratton. (2000). On the Condition Numbers Associated with the Polar Factorization of a Matrix, *Numer. Lin. Alg. Appl.* **7**, pp. 337–354.

K. L. Chu, J. Isotalo, S. Puntanen, and G. P. H. Styan. (2005). The Efficiency Factorization Multiplier for the Watson Efficiency in Partitioned Linear Models: Some Examples and a Literature Review, *Res. Lett. Inf. Math. Sci.* **8**, pp. 165–187.

C. Chui. (1992). *An Introduction to Wavelets*, (Academic Press).

J. Clauser, M. Horne, A. Shimony and R. Holt. (1969). Proposed Experiment to Test Local Hidden-Variable Theories, *Phys. Rev. Lett.* **23**, pp. 880–884.

Y. Coudene. (2006). Pictures of Hyperbolic Dynamical Systems, *Notices of the AMS* **53**(1), pp. 8–13.

D. Cudia. (1964). The Geometry of Banach Spaces. Smoothness, *Trans. Amer. Math. Soc.* **110**, pp. 284–314.

N. Cuntoor and R. Chellappa. (2006). Key Frame-Based Activity Representation Using Antieigenvalues, in *Computer Vision–ACCV 2006*, Narayanan P., Nayar S., and Shum H. Y. (eds.), Springer Lecture Notes in Computer Science, Vol. 3852/2006, (Springer), pp. 499–508.

I. Daubechies. (1990). The Wavelet Transform, Time-Frequency Localization and Signal Analysis, *IEEE Trans. Information Theory* **36**, pp. 961–1005.

I. Daubechies. (1992). *Ten Lectures on Wavelets*, (SIAM Publications, Philadelphia).

J. Dauxois, G. Nkiet and Y. Romain. (2004). Canonical Analysis Relative to a Closed Subspace, *Lin. Alg. in Applic.* **388**, pp. 119–145.

F. Delbaen and W. Schachermayer. (2006). *The Mathematics of Arbitrage*, (Springer, Berlin).

J. Dorroh. (1966). Contraction Semigroups in a Function Space, *Pacific J. Math.* **19**, pp. 35–38.

R. Duffin and A. Schaeffer. (1952). A Class of Nonharmonic Fourier Series, *Trans. Amer. Math. Soc.* **72**, pp. 341–366.

N. Dunford and J. Schwartz. (1971). *Linear Operators III: Spectral Operators*, (Wiley & Sons, New York).

J. Durbin and G. S. Watson. (1950). Testing for Serial Correlation in Least Squares Regression. I, *Biometrica* **37**, pp. 409–428.

J. Durbin and G. S. Watson. (1951). Testing for Serial Correlation in Least Squares Regression. II, *Biometrica* **38**, pp. 159–177.

M. Eaton. (1976). A Maximization Problem and its Application to Canonical Correlation, *J. Multivar. Anal.* **6**, pp. 422–425.

A. Einstein, B. Podolsky and N. Rosen. (1935). Can Quantum Mechanical Description of Reality be Considered Complete?, *Phys. Rev.* **47**, pp. 777–780.

K. Fan. (1966). Some Matrix Inequalities, *Abh. Math. Sem. Univ. Hamburg* **29**, pp. 185–196.

H. Föllmer and A. Schied. (2004). *Stochastic Finance*, (de Gruyter, Berlin).

G. Golub and C. F. Van Loan. (1987). *Matrix Computations*, (Johns Hopkins Univ. Press).

T. Goodman, S. Lee and W. Tang. (1993). Wavelets in Wandering Subspaces, *Trans. Amer. Math. Soc.* **338**, pp. 639-654.

G. Gorton. (2008). *The Panic of 2007*, report prepared for the Federal Reserve Bank of Kansas City, Jackson Hole Conference, August 2008. Available online (90 pp.).

W. Greub and W. Rheinboldt. (1959). On a Generalization of an Inequality of L. V. Kantorovich, *Proc. Amer. Math. Soc.* **10**, pp. 407–415.

K. Gustafson. (1973a). Some Essentially Selfadjoint Dirac Operators with Spherically Symmetric Potentials, *Israel J. Math.* **14**, pp. 63–75 (with P. Rejto).

K. Gustafson. (1981a). Partial Inner Product Spaces and Semi-Inner Product Spaces, *Adv. Math.* **41**, pp. 281–300 (with J. P. Antoine).

K. Gustafson. (1981b). Timestep Control for the Numerical Solutions of Initial-Boundary Value Problems, *Quantum Mechanics in Mathematics, Chemistry, and*

Physics, Gustafson K. and Reinhardt W. (eds.), (Plenum Press, NY), pp. 407–414 (with J. Gary and H. Tadjeran).

K. Gustafson. (1982a). Exact Solutions and Ignition Parameters in the Arrhenius Conduction Theory of Gaseous Thermal Explosion, *ZAMP* **33**, pp. 391–405 (with B. Eaton).

K. Gustafson. (1983b). Divergence-Free Bases for Finite Element Schemes in Hydrodynamics, *SIAM J. Numer. Anal.* **20**, pp. 697–721. (with R. Hartman).

K. Gustafson. (1985a). A New Method for Computing Solenoidal Vector Fields on Arbitrary Regions, *Int. J. Numer. Meth. Fluid* **5**, pp. 763–783 (with K. Halasi, D. P. Young).

K. Gustafson. (1987a). Kolmogorov Systems and Haar Systems, *Colloq. Math. Soc. Janos Bolyai* **49**, pp. 401–416 (with R. K. Goodrich).

K. Gustafson. (1988a). Multigrid Localization and Multigrid Grid Generation for the Computation of Vortex Structures and Dynamics of Flows in Cavities and About Airfoils, in *Multigrid Methods*, McCormick S., (ed.), (Dekker), pp. 229–250 (with R. Leben).

K. Gustafson. (1990a). *Vortex Methods and Vortex Motion*, (SIAM Publications, Philadelphia) (with J. Sethian).

K. Gustafson. (1990b). Counting the Number of Solutions in Combustion and Reactive Flow Problems, *J. Appl. Math. Phys.* **41**, pp. 558–578 (with E. Ash, B. Eaton).

K. Gustafson. (1991b). *Computational Physics*, (North Holland, Amsterdam) (with W. Wyss).

K. Gustafson. (1991c). Preconditioned Conjugate Gradient and Finite Element Methods for Massively Data-Parallel Architectures, *Comput. Phys. Commun.* **65**, pp. 253–267 (with N. Sobh).

K. Gustafson. (1992a). Lift and Thrust Generation by an Airfoil in Hover Modes, *Comput. Fluid Dynam. J.* **1**, pp. 47–57 (with R. Leben, J. McArthur).

K. Gustafson. (1994d). *Risk in Portfolio Management*, Asea Brown Boveri (ABB) Internal Report, p. 18.

K. Gustafson. (1998h). Haar Wavelets and Differential Equations, *Differential Equations* **34**, pp. 829–832 (with I. Antoniou).

K. Gustafson. (1999f). *Introduction to Partial Differential Equations and Hilbert Space Methods*, 3rd edn. (Revised), (Dover Publications, Mineola, NY).

K. Gustafson. (1999g). Wavelets and Stochastic Processes, *Math. Comput. Simulat.* **49**, pp. 81–104 (with I. Antoniou).

K. Gustafson. (1999h). Parallel Computing Forty Years Ago, *Math. Comput. Simulat.* **51**, pp. 47–62.

K. Gustafson. (2000e). The Time Operator of Wavelets, *Chaos, Solitons, and Fractals* **11**, pp. 443–452 (with I. Antoniou).

K. Gustafson. (2000f). *A 9th Derivation of the Black–Scholes Equation*, preprint.

K. Gustafson. (2002c). Distinguishing Discretization and Discrete Dynamics, with Application to Ecology, Machine Learning, and Atomic Physics, in *Structure and Dynamics of Nonlinear Wave Phenomena*, Tanaka M. (ed.), RIMS Kokyuroku, Vol. 1271, Kyoto, Japan.

K. Gustafson. (2002d). Time–Space Dilations and Stochastic–Deterministic Dynamics, in *Between Chance and Choice*, Atmanspacher H. and Bishop R. (eds.), (Imprint Academic, UK), pp. 115–148.

K. Gustafson. (2007c). Wavelets and Expectations: A Different Path to Wavelets, in *Harmonic, Wavelet and p-adic Analysis*, Chuong N., Egorov Y., Khrennikov A., Meyer Y. and Mumford D., (eds.), (World Scientific), pp. 5–22.

K. Gustafson. (2010f). Experiences and Insights in Mathematical Finance, *J. Shanghai Finance Univ.* No. 4, pp. 35–41 (In Chinese).

K. Gustafson. (2011d). Normal Degree, Revisited. In preparation.

K. Gustafson. (2011e). *The Crossing of Heaven: Memoirs of a Mathematician*, (Springer-Heidelberg), publication date 28 Dec. 2011.

P. Halmos. (1967). *A Hilbert Space Problem Book*, (D. Van Nostrand, Princeton, NJ).

E. Hille and R. S. Phillips. (1957). *Functional Analysis and Semigroups, Colloq. Publ. Amer. Math. Soc.*, (AMS Providence, RI).

R. Horn and C. Johnson. (1985). *Matrix Analysis*, (Cambridge).

G. Jaeger and K. Ann. (2008). Entanglement Sudden Dealth in a Qubit–Qutrit System. *Phys. Lett. A* **372**, pp. 579–583.

N. Jewell, P. Bloomfield and F. Bartmann. (1983). Canonical Correlations of Past and Future for Time Series: Bounds and Computation, *Ann. Statist.* **11**, pp. 848–855.

S. Johnson. (2009). The Quiet Coup, *The Atlantic Magazine*, May issue.

I. Kaporin. (1990). New Convergence Results and Preconditioning Strategies for the Conjugate Gradient Method, *Numer Lin. Alg. with Applic.* **1**, 179–210.

T. Kato. (1976). *Perturbation Theory for Linear Operators*, (Springer, Berlin), 13 pp.

M. Knill and R. Laflamme. (1997). Theory of Quantum Error-Correcting Codes. *Phys. Rev. A* **55**, pp. 900–911.

M. Kus and K. Zyczkowski. (2001). Geometry of Entangled States, *Phys. Rev. A* **63**, 032307, 13 pp.

N. Levan and C. Kubrusly. (2003). A Wavelet "Time-Shift-Detail" Decomposition, *Math. Comput. Simulat.* **63**, pp. 73–78.

N. Levan and C. Kubrusly. (2004). Time-Shifts Generalized Multiresolution Analysis over Dyadic-Scaling Reducing Subspaces, *Int. J. Wavelets Multiresolut. Inf. Process.* **2**, pp. 1–12.

M. Lewis. (2010). *The Big Short*, (Norton).

J. Liesen and P. Saylor. (2005). Orthonormal Hessenberg Reduction and Orthogonal Krylov Subspace Bases, *SIAM J. Numer Anal.* **42**, pp. 2148–2158.

D. Luenberger. (1973). *Introduction to Linear and Nonlinear Programming*, (Addison-Wesley Publ. Co.).

G. Lumer and R. S. Phillips. (1961). Dissipative Operators in a Banach Space, *Pacific J. Math.* **11**, pp. 679–698.

D. MacKenzie and Y. Millo. (2003). Constructing a Market, Performing Theory: The Historical Sociology of a Financial Derivatives Exchange, *Amer. J. Sociol.* **109**, pp. 107–145.

W. Magnus. (1974). *Noneuclidean Tesselations and their Groups*, (Academic Press, New York).

A. Marshall and I. Olkin. (1990). Matrix Versions of the Cauchy and Kantorovich Inequalities, *Aequationes Math.* **40**, pp. 89–93.

C. McCarthy. (1967). C_p, *Israel J. Math* **5**, pp. 249–271.

C. Mendl and M. Wolf. (2009). United Quantum Channels — Convex Structure and Revivals of Birkhoff's Theorem. *Commun. Math. Phys.* **289**, pp. 1057–1086.

N. D. Mermin. (2007). *Quantum Computer Science: An Introduction*, (Cambridge Univ. Press, Cambridge).

R. C. Merton. (1990). *Continuous Time Finance*. (Basil Blackwell, Oxford, UK).

C. Meyer. (2000). *Matrix Analysis and Applied Linear Algebra*, (SIAM Publications, Philadelphia).

B. S. Nagy and C. Foias. (1970). *Harmonic Analysis of Operators in Hilbert Space*, (North Holland).

E. Nelson. (1964). Feynman Integrals and the Schrodinger Equation, *J. Math. Phys.* **5**, pp. 332–343.

M. Nielsen and I. Chuang. (2000). *Quantum Computation and Quantum Information*, (Cambridge Press, Cambridge).

I. Olkin. (1981). Range Restrictions for Product–Moment Correlation Matrices, *Psychometrika* **46**, pp. 469–472.

R. Ortega and M. Santander. (2002). Trigonometry of "complex Hermitian"-Type Homogeneous Symmetric Spaces, *J. Phys. A: Math. Gen.* **35**, pp. 7877–7917.

S. Patterson. (2010). *The Quants*. (Random House).

R. Penrose. (2005). *The Road to Reality*, (Alfred Knopf, New York).

M. Pourahmadi. (2001). *Foundations of Time Series Analysis and Prediction Theory*, (Wiley, New York).

C. R. Rao and C. V. Rao. (1987). Stationary Values of the Product of Two Rayleigh Quotients: Homologous Canonical Correlations, *Sankhya* **49B**, pp. 113–125.

B. Schwarzschild. (1999). At Last We have an Undisputed Observation of "Direct" CP Violation in Kaon Decay. *Physics Today*, May, 1999, pp. 17–19.

B. Schwarzschild. (2003). Antineutrinos from Distant Reactors Simulate the Disappearance of Solar Neutrinos, *Physics Today*, March, pp. 14–16.

J. Stampfli. (1970). The Norm of a Derivation, *Pacific J. Math.* **33**, pp. 737–747.

G. Strang. (1962). Eigenvalues of Jordan Products, *Amer. Math. Monthly* **69**, pp. 37–40.

N. Touzi. (1999). American Option Exercise Boundary when the Volatility Changes Randomly, *Appl. Math. Optimiz.* **39**, pp. 411–422.

G. 't Hooft. (2002). Determinism Beneath Quantum Mechanics, ArXiv: quant-ph/0212095v1.

R. Varga. (1962). *Matrix Iterative Analysis*, (Prentice Hall).

J. Von Neumann. (1941). Distribution of the Ratio of Mean Square Successive Difference to the Variance, *Annals of Math. Stat.* **12**, pp. 367–395.

J. von Neumann and O. Morgenstern. (1944). *Theory of Games and Economic Behavior*, (Princeton Univ. Press).

J. Von Neumann, R. Kent, H. Bellinson and B. Hart. (1941). The Mean Square Successive Difference, *Annals of Math. Stat.* **12**, pp. 153–162.

S. G. Wang and S. C. Chow. (1994). *Advanced Linear Models*, (Marcel Dekker, New York).

S. Wang and W. Ip. (2000). A Matrix Version of the Wielandt Inequality and its Applications to Statistics, *Lin. Alg. and Applic.* **296**, pp. 171–181.

E. Wigner. (1970). On Hidden Variables and Quantum Mechanical Probabilities, *Amer. J. Phys.* **38**, pp. 1005–1009.

P. Wilmott. (1998). *Derivatives: The Theory and Practice of Financial Engineering*, (John Wiley & Sons, Chichester).

H. Yanai, S. Putanen, K. Ito and H. Ishii. (2008). Some Extensions on the Ranges of a Correlation Matrix and its Inverse Matrix, preprint, 14 pp.

K. Yosida. (1968). *Functional Analysis*, (Springer, Berlin).

D. M. Young. (1971). *Iterative Solutions of Large Linear Systems*, (Academic Press).

R. Zvan, P. Forsyth and K. Vetzal. (1998). Penalty Methods for American Options with Stochastic Volatility. *J. Comput. Appl. Math.* **91**, pp. 199–218.

Index

$\phi(A)$, 2

absolute condition number, 64
Accardi, L., 123
accretive operator, 5
additive Schwarz, 62
ADI, 60
Alfred Haar Memorial Conference, 69
Alfred Haar's 1910 dissertation, 72
angles, 1, 2, 6, 7, 58, 186
Antieigenmatrices, 115
antieigenmatrix, 65
antieigenvalue, 1, 2, 6, 33, 34
antieigenvector, 1, 6, 7, 30, 31, 38, 40
Antieigenvector Pair Angle, 44, 85
Antieigenvector-Breaking, 143
antieigenvectors, 7, 30, 31, 33, 39, 84
Antoine, J.P., 25
Ash, E., 54
Asplund, E., 49

BA, 4, 14, 63
Basis Trigonometry, 78
Bell's inequalities, 123
Bell's inequality, 153
Berlin, 164
bid–ask gap, 159
Black–Scholes, 23
Black–Scholes PDE, 157
Bloch sphere, 154
Breckenridge, 66
Brussels, 69
Budapest, 69

Calvert, B., 19
Canonical correlations, 105, 114
Chaitin-Chatelin, F., 64
coarse-graining, 71
combinatorial viewpoint, 39
computational experience, 53
Computational Fluid Dynamics, 157
condition number, 43, 63, 188
conjugate gradient, 67
contraction semigroup, 2, 3, 5, 16, 17, 18, 21
    completely nonunitary, 88
Convexity, 26
$\cos \phi(A)$, 4

Davenport, J., 159
Davis, C., 13, 50, 184
De Moivre's theorem, 83
Decoherence, 146
Dirichlet boundary conditions, 82
dispersiveness, 22
dissipativeness, 22
Domain Decomposition, 61
Dorroh, J., 18
Dubrovnik, ix, 55
Dubrovnik Workshop, 53
Durbin–Bloomfield–Watson–Knott lower bound, 92

Eaton, B., 54
eigenvalue, 1, 2
eigenvectors, 1
Elementary Particles, 142
Entanglement, 144

Entropy, 145
EPR, 125
Euler Equation, 33, 36, 41, 101
Extended Operator Trigonometry, 47

financial crashes, 163
Financial Instruments, 155
Fréchet derivative, 64
frame, 74, 89
frame operator, 74
    maximal turning angle, 75
Frobenius norm, 64
Fundamental Theorem of Linear
    Algebra, 203

general spinor, 141
golden mean, 202
Gradient Descent, 54
Gram matrix, 46
Gram matrix determinant, 47, 198
graph norm, 23
Green's function, 16
Greub–Rheinboldt bounds, 21

Haar Basis, 73
Haar wavelet, 73, 89
Halasi, K., 54
Hartman, R., 54
heat equation, 16
Hess, P., 189
Higher Antieigenvalues, 39
Hille–Yosida, 16
Hotelling correlation, 108
Hydrogen, 18

imaginary operator trigonometry, 186
Inefficiency Equation, 101
inequality–equalities, 129, 153, 191
Infinitesimal Generators, 17
Initial Value Problems, 2, 16
interaction-antieigenvalues, 187
intra-angle, 84
Irreversibility, 86
Ito, K., 101, 118
IWMS 2005, 91

Jackson Hole, Wyoming, 70
Jordan canonical form, 213

Kantorovich, L., 50, 53, 65
Kantorovich–Wielandt inequality, 55
Kaporin, I., 188
Kato, T., 17
Khatree, R., 13
Kolmogorov systems, 70, 88
Kolmogorovian probability, 124
Krein, M.G., 13, 46, 47, 49

Lagrange multiplier methods, 42
Lagrangian variational methods, 101
Laplacian, 23, 58, 82
Laplacian operator, 16
Leben, R., 54
Les Treilles in Provence, 123
light cone, 135
Littlewood–Paley basis, 79
Los Angeles, ix, 27, 49
Lyapunov Stability, 85

Markov process, 22
matrix products, 4
matrix statistics, 34
Maybee, J., 27
McArthur, J., 54
McCarthy, C., 20
Min-Max, 27
Min-Max Theorem, 5, 27, 31, 104, 137
Minimum Residual, 56
Mirman, B., 13, 50
Model Problem, 58
Morse theory, 45
most-turned vectors, 6
Multiplicative Perturbation, 18, 86
multiplicatively perturb, 3
multiresolution analyses, 71

NAPA Preconditioning Conference, 64
Navier elasticity, 55
Nelson, E., 17
Netherlands, 66
New Zealand, 91
noncommuting, 4

total operator trigonometry, 186
"total" turning angle, 132
Triangle Inequality, 46
Trigonometry of efficiency, 93
Triple Commute, 63
twisting vector, 80
Twistors, 135, 138
twistvectors, 49

University of Minnesota, 20, 21

Volatility, 177
Von Neumann, J., 98, 118

wandering subspace, 69, 88
wavelet, 70, 88
Wavelet Reconstruction, 76
Wielandt, H., 13, 49
Wigner, 125
Wigner inequality–equality, 153
Wisconsin, 49, 55

Young, D.M., 58, 60
Young, D.P., 54

Zwahlen, B., 19

normal operators, 184
numerical range, 4, 50

oblique projections, 14
one-parameter unitary group, 86
operator angle, 58
operator trigonometry, 2
operator turning, 1
optimal, 57, 58
Optimum, 31, 32
ordinary differential equations, 23
orthonormal basis, 75, 79

Pascal's triangle, 83
Penrose, R., 124
Poisson–Neumann boundary value
    problem, 101
polar form, 48
Portfolio, 175
positivity of matrix products, 20
Preconditioning, 63
Prediction Theory, 112
Ptak, V., 49
Puntanen, S., 117

quadratic equation, 43
Quantos, 167
Quantum Computing, 132
Quantum Fidelity, 149
Quantum Spin Identities, 129
Quantum States, 144
quark mixing, 143

Rao, C.R., 13, 91, 105, 117
Rao, D., 13, 45, 46, 49
Rayleigh eigenvalue, 90
Rayleigh quotient, 14, 105
Rayleigh variational quotients, 33
Rayleigh–Ritz theory, 26
re-inner-product, 184
reducing subspace, 41
Rejto, P., 20
Rellich, F., 17
Richardson iteration, 74, 77
Richardson Relaxation, 57
Risk axioms, 160

Sato, K., 19, 21
Schneider, H., 49, 55
Schrödinger equation, 86
Schrodinger, E., 17, 18
Schwarz's inequality, 14, 32
second antieigenvector, 7
Seddighin, M., 13, 45, 184
Seguin, T., 159
semigroup, 2, 23, 86
Seneta, E., 91
Shanghai Finance University, 16
Shisha–Mond inequality, 190, 20
$\sin \phi(A)$, 5, 8, 9, 194
singular value, 48, 64
skew antieigenvalues, 44
slant antieigenvalues, 186
Sobh, N., 53
Solvay Institute, 69, 123
Solvers, 56
SOR method, 58
spacetime angle, 135
spectral radius, 45, 59
spectral representation, 14
spectral theory, 1
spin model, 124
Spread, 172
Stampfli, J., 13, 21
Statistical Efficiency, 93
Statistics, 91
Statistics Inequalities, 107
Stimmel, E., 14
Stone–Von Neumann Theorem, 89
Strang, G., 13, 21
Sturm–Liouville boundary value
    problem, 81
Styan, G., 117
suddenly incomplete market, 164
symmetric antieigenvalue, 186
"symplectic" representation, 197

Tadjeran, H., 54
't Hooft, G., 200
Thorp, E., 182
Time operator, 69, 72
Toeplitz–Hausdorff Theorem, 50
Toronto, 50